NATURAL DISASTER RESEARCH, PREDICTION AND MITIGATION

NATIONAL RESPONSE AND DISASTER RECOVERY FRAMEWORKS

NATURAL DISASTER RESEARCH, PREDICTION AND MITIGATION

Additional books in this series can be found on Nova's website
under the Series tab.

Additional e-books in this series can be found on Nova's website
under the e-book tab.

NATURAL DISASTER RESEARCH, PREDICTION AND MITIGATION

NATIONAL RESPONSE AND DISASTER RECOVERY FRAMEWORKS

DANIELLA BECKETT
EDITOR

nova publishers
New York

For permission to use material from this book please contact us:
Telephone 631-231-7269; Fax 631-231-8175
Web Site: http://www.novapublishers.com

Library of Congress Cataloging-in-Publication Data

ISBN: 978-1-63321-892-5

Published by Nova Science Publishers, Inc. † New York

CONTENTS

PREFACE

The structures, roles, and responsibilities described in this Framework can be partially or fully implemented in the context of a threat or hazard, in anticipation of a significant event, or in response to an incident. Selective implementation of National Response Framework structures and procedures allows for a scaled response, delivery of the specific resources and capabilities, and a level of coordination appropriate to each incident. This book discusses the National Response Framework, as well as the National Disaster Recovery Framework in full detail.

Chapter 1 – The National Response Framework is a guide to how the Nation responds to all types of disasters and emergencies. It is built on scalable, flexible, and adaptable concepts identified in the National Incident Management System to align key roles and responsibilities across the Nation. This Framework describes specific authorities and best practices for managing incidents that range from the serious but purely local to large-scale terrorist attacks or catastrophic natural disasters. The National Response Framework describes the principles, roles and responsibilities, and coordinating structures for delivering the core capabilities required to respond to an incident and further describes how response efforts integrate with those of the other mission areas.

Chapter 2 –The National Disaster Recovery Framework is a guide that enables effective recovery support to disaster-impacted States, Tribes, Territorial and local jurisdictions. It provides a flexible structure that enables disaster recovery managers to operate in a unified and collaborative manner. It also focuses on how best to restore, redevelop and revitalize the health, social, economic, natural and environmental fabric of the community and build a more resilient Nation.

In: National Response and Disaster Recovery Frameworks ISBN: 978-1-63321-892-5
Editor: Daniella Beckett

Chapter 1

NATIONAL RESPONSE FRAMEWORK, SECOND EDITION*

U.S. Department of Homeland Security

EXECUTIVE SUMMARY

The National Response Framework is a guide to how the Nation responds to all types of disasters and emergencies. It is built on scalable, flexible, and adaptable concepts identified in the National Incident Management System to align key roles and responsibilities across the Nation. This Framework describes specific authorities and best practices for managing incidents that range from the serious but purely local to large-scale terrorist attacks or catastrophic natural disasters. The National Response Framework describes the principles, roles and responsibilities, and coordinating structures for delivering the core capabilities required to respond to an incident and further describes how response efforts integrate with those of the other mission areas. **This Framework is always in effect, and elements can be implemented at any time.** The structures, roles, and responsibilities described in this Framework can be partially or fully implemented in the context of a threat or hazard, in anticipation of a significant event, or in response to an incident. Selective implementation of National Response Framework structures and procedures allows for a scaled response, delivery of the specific resources and capabilities, and a level of coordination appropriate to each incident.

The Response mission area focuses on ensuring that the Nation is able to respond effectively to all types of incidents that range from those that are adequately handled with local assets to those of catastrophic proportion that require marshaling the capabilities of the entire Nation. The objectives of the Response mission area define the capabilities necessary to save lives, protect property and the environment, meet basic human needs, stabilize the incident, restore basic services and community functionality, and establish a safe and secure environment moving toward the transition to recovery.[1] The Response mission area includes 14 core capabilities: planning, public information and warning, operational coordination,

* This is an edited, reformatted and augmented version of a document issued May 2013.

critical transportation, environmental response/health and safety, fatality management services, infrastructure systems, mass care services, mass search and rescue operations, on-scene security and protection, operational communications, public and private services and resources, public health and medical services, and situational assessment.

The priorities of response are to save lives, protect property and the environment, stabilize the incident and provide for basic human needs. The following principles establish fundamental doctrine for the Response mission area: engaged partnership, tiered response, scalable, flexible, and adaptable operational capabilities, unity of effort through unified command, and readiness to act.

Scalable, flexible, and adaptable coordinating structures are essential in aligning the key roles and responsibilities to deliver the Response mission area's core capabilities. The flexibility of such structures helps ensure that communities across the country can organize response efforts to address a variety of risks based on their unique needs, capabilities, demographics, governing structures, and non-traditional partners. This Framework is not based on a one-size-fits-all organizational construct, but instead acknowledges the concept of tiered response which emphasizes that response to incidents should be handled at the lowest jurisdictional level capable of handling the mission.

In implementing the National Response Framework to build national preparedness, partners are encouraged to develop a shared understanding of broad-level strategic implications as they make critical decisions in building future capacity and capability. The whole community should be engaged in examining and implementing the strategy and doctrine contained in this Framework, considering both current and future requirements in the process.

INTRODUCTION

The National Response Framework (NRF) is an essential component of the National Preparedness System mandated in Presidential Policy Directive (PPD) 8: National Preparedness. PPD-8 is aimed at strengthening the security and resilience of the United States through systematic preparation for the threats that pose the greatest risk to the security of the Nation. PPD-8 defines five mission areas—Prevention, Protection, Mitigation, Response, and Recovery—and mandates the development of a series of policy and planning documents to explain and guide the Nation's collective approach to ensuring and enhancing national preparedness. The NRF sets the doctrine for how the Nation builds, sustains, and delivers the response core capabilities identified in the National Preparedness Goal (the Goal). The Goal establishes the capabilities and outcomes the Nation must accomplish across all five mission areas in order to be secure and resilient

Prevention: The capabilities necessary to avoid, prevent, or stop a threatened or actual act of terrorism. As defined by PPD-8, the term "prevention" refers to preventing imminent threats.

Protection: The capabilities necessary to secure the homeland against acts of terrorism and manmade or natural disasters.

Mitigation: The capabilities necessary to reduce loss of life and property by lessening the impact of disasters.

Response: The capabilities necessary to save lives, protect property and the environment, and meet basic human needs after an incident has occurred.

Recovery: The capabilities necessary to assist communities affected by an incident to recover effectively.

Framework Purpose and Organization

The NRF is a guide to how the Nation responds to all types of disasters and emergencies. It is built on scalable, flexible, and adaptable concepts identified in the National Incident Management System (NIMS)[2] to align key roles and responsibilities across the Nation. The NRF describes specific authorities and best practices for managing incidents that range from the serious but purely local to large-scale terrorist attacks or catastrophic[3] natural disasters.

This document supersedes the NRF that was issued in January 2008. It becomes effective 60 days after publication.

The term "response," as used in the NRF, includes actions to save lives, protect property and the environment, stabilize communities, and meet basic human needs following an incident. Response also includes the execution of emergency plans and actions to support short-term recovery. The NRF describes doctrine for managing any type of disaster or emergency regardless of scale, scope, and complexity. This Framework explains common response disciplines and processes that have been developed at all levels of government (local, state, tribal, territorial, insular area,[4] and Federal) and have matured over time.

To support the Goal, the objectives of the NRF are to:

- Describe scalable, flexible, and adaptable coordinating structures, as well as key roles and responsibilities for integrating capabilities across the whole community,[5] to support the efforts of local, state, tribal, territorial, insular area, and Federal governments in responding to actual and potential incidents
- Describe, across the whole community, the steps needed to prepare for delivering the response core capabilities
- Foster integration and coordination of activities within the Response mission area
- Outline how the Response mission area relates to the other mission areas, as well as the relationship between the Response core capabilities and the core capabilities in other mission areas
- Provide guidance through doctrine and establish the foundation for the development of the supplemental Response Federal Interagency Operational Plan (FIOP).

The NRF is composed of a base document, Emergency Support Function (ESF) Annexes, Support Annexes, and Incident Annexes (see Figure 1). The annexes provide detailed information to assist with the implementation of the NRF.

- **ESF Annexes** describe the Federal coordinating structures that group resources and capabilities into functional areas that are most frequently needed in a national response.
- **Support Annexes** describe the essential supporting processes and considerations that are most common to the majority of incidents.
- **Incident Annexes** describe the unique response aspects of incident categories.

Evolution of the Framework

This NRF is currently the most mature of the National Planning Frameworks because it builds on 20 years of Federal response guidance. The Federal Response Plan published in 1992 focused largely on Federal roles and responsibilities. The establishment of the Department of Homeland Security (DHS) and the emphasis on the development and implementation of common incident management and response principles led to the development of the National Response Plan (NRP) in 2004. The NRP broke new ground by integrating all levels of government, the private sector, and nongovernmental organizations (NGOs) into a common incident management framework. In 2008, the NRP was superseded by the first NRF, which streamlined the guidance and integrated lessons learned from Hurricane Katrina and other incidents.

This NRF reiterates the principles and concepts of the 2008 version of the NRF and implements the new requirements and terminology of PPD-8. By fostering a holistic approach to response, this NRF emphasizes the need for the involvement of the whole community. Along with the National Planning Frameworks for other mission areas, this document now describes the all-important integration and inter-relationships among the mission areas of Prevention, Protection, Mitigation, Response, and Recovery.

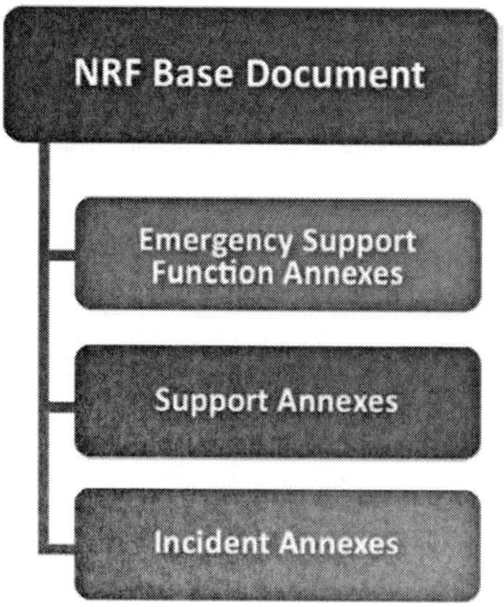

Figure 1. Organization of the NRF.

Relationship to NIMS

The response protocols and structures described in the NRF align with NIMS. NIMS provides the incident management basis for the NRF and defines standard command and management structures. Standardizing national response doctrine on NIMS provides a consistent, nationwide template to enable the whole community to work together to prevent, protect against, mitigate, respond to, and recover from the effects of incidents regardless of cause, size, location, or complexity.

All of the components of the NIMS—including preparedness, communications and information management, resource management, and command and management—support response. The NIMS concepts of multi-agency coordination and unified command are described in the command and management component of NIMS. These two concepts are essential to effective response operations because they address the importance of: (1) developing a single set of objectives; (2) using a collective, strategic approach; (3) improving information flow and coordination; (4) creating a common understanding of joint priorities and limitations; (5) ensuring that no agency's legal authorities are compromised or neglected; and (6) optimizing the combined efforts of all participants under a single plan.

Intended Audience

The NRF is intended to be used by the whole community. This all-inclusive concept focuses efforts and enables a full range of stakeholders—individuals, families, communities, the private and nonprofit sectors, faith-based organizations, and local, state, tribal, territorial, insular area, and Federal governments—to participate in national preparedness activities and to be full partners in incident response. Government resources alone cannot meet all the needs of those affected by major disasters. All elements of the community must be activated, engaged, and integrated to respond to a major or catastrophic incident.

Engaging the whole community is essential to the Nation's success in achieving resilience and national preparedness. Individual and community preparedness is a key component to this objective. By providing equal access to acquire and use the necessary knowledge and skills, the whole community contributes to and benefits from national preparedness. This includes children; individuals with disabilities and others with access and functional needs;[6] those from religious, racial and ethnically diverse backgrounds; and people with limited English proficiency. Their contributions must be integrated into preparedness efforts, and their needs must be incorporated into planning for and delivering the response core capabilities as defined in the Goal.[7]

Although the NRF is intended to provide guidance for the whole community, it focuses especially on the needs of those who are involved in delivering and applying the response core capabilities defined in the National Preparedness Goal. This includes emergency management practitioners, community leaders, and government officials who must collectively understand and assess the needs of their respective communities and organizations and determine the best ways to organize and strengthen their resiliency.

SCOPE

The NRF describes structures for implementing nationwide response policy and operational coordination for all types of domestic incidents.[8] This section describes the scope of the Response mission area, the guiding principles of response doctrine and their application, and how risk informs response planning.

The Response mission area focuses on ensuring that the Nation is able to respond effectively to all types of incidents that range from those that are adequately handled with local assets to those of catastrophic proportion that require marshaling the capabilities of the entire Nation. The objectives of the Response mission area define the capabilities necessary to save lives, protect property and the environment, meet basic human needs, stabilize the incident, restore basic services and community functionality, and establish a safe and secure environment moving toward the transition to recovery.[9]

The NRF describes the principles, roles and responsibilities, and coordinating structures for delivering the core capabilities required to respond to an incident and further describes how response efforts integrate with those of the other mission areas. **The NRF is always in effect, and elements can be implemented at any time.** The structures, roles, and responsibilities described in the NRF can be partially or fully implemented in the context of a threat or hazard, in anticipation of a significant event, or in response to an incident. Selective implementation of NRF structures and procedures allows for a scaled response, delivery of the specific resources and capabilities, and a level of coordination appropriate to each incident.

In this Framework, the term incident includes actual or potential emergencies and disasters resulting from all types of threats and hazards, ranging from accidents and natural disasters to cyber intrusions and terrorist attacks. The NRF's structures and procedures address incidents where Federal support to local, state, tribal, territorial, and insular area governments is coordinated under the Robert T. Stafford Disaster Relief and Emergency Assistance Act (Stafford Act), as well as incidents where Federal departments and agencies exercise other authorities and responsibilities.

Nothing in the NRF is intended to alter or impede the ability of any local, state, tribal, territorial, insular area, or Federal government department or agency to carry out its authorities or meet its responsibilities under applicable laws, executive orders, and directives.

Guiding Principles

The priorities of response are to save lives, protect property and the environment, stabilize the incident and provide for basic human needs. The following principles establish fundamental doctrine for the Response mission area: (1) engaged partnership, (2) tiered response, (3) scalable, flexible, and adaptable operational capabilities, (4) unity of effort through unified command, and (5) readiness to act. These principles are rooted in the Federal system and the Constitution's division of responsibilities between state and Federal governments. These principles reflect the history of emergency management and the distilled wisdom of responders and leaders across the whole community.

Engaged Partnership

Effective partnership relies on engaging all elements of the whole community, as well as international partners in some cases. This also includes survivors who may require assistance and who may also be resources to support community response and recovery.

Those who lead emergency response efforts must communicate and support engagement with the whole community by developing shared goals and aligning capabilities to reduce the risk of any jurisdiction being overwhelmed in times of crisis. Layered, mutually supporting capabilities of individuals, communities, the private sector, NGOs, and governments at all levels allow for coordinated planning in times of calm and effective response in times of crisis. Engaged partnership and coalition building includes ongoing clear, consistent, effective,[10] and culturally appropriate communication and shared situational awareness about an incident to ensure an appropriate response.

Tiered Response

Most incidents begin and end locally and are managed at the local level. These incidents typically require a unified response from local agencies, the private sector, and NGOs. Some may require additional support from neighboring jurisdictions or state governments. A smaller number of incidents require Federal support or are led by the Federal Government.[11] National response protocols are structured to provide tiered levels of support when additional resources or capabilities are needed.

Scalable, Flexible, and Adaptable Operational Capabilities

As incidents change in size, scope, and complexity, response efforts must adapt to meet evolving requirements. The number, type, and sources of resources must be able to expand rapidly to meet the changing needs associated with a given incident and its cascading effects. As needs grow and change, response processes must remain nimble and adaptable. The structures and processes described in the NRF must be able to surge resources from the whole community. As incidents stabilize, response efforts must be flexible to support the transition from response to recovery.

Unity of Effort through Unified Command

Effective, unified command is indispensable to response activities and requires a clear understanding of the roles and responsibilities of all participating organizations.[12] The Incident Command System (ICS), a component of NIMS, is an important element in ensuring interoperability across multi-jurisdictional or multi-agency incident management activities. Unified command, a central tenet of ICS, enables organizations with jurisdictional authority or functional responsibility for an incident to support each other through the use of mutually developed incident objectives. Each participating agency maintains its own authority, responsibility, and accountability.

Readiness to Act

Effective response requires a readiness to act that is balanced with an understanding of the risks and hazards responders face. From individuals, families, and communities to local, state, tribal, territorial, insular area, and Federal governments, national response depends on the ability to act decisively. A forward-leaning posture is imperative for incidents that may expand rapidly in size, scope, or complexity, as well as incidents that occur without warning. Decisive action is often required to save lives and protect property and the environment. Although some risk to responders may be unavoidable, all response personnel are responsible for anticipating and managing risk through proper planning, organizing, equipping, training, and exercising. Effective response relies on disciplined processes, procedures, and systems to communicate timely, accurate, and accessible information about an incident's cause, size, and current status to the public, responders, and other stakeholders.

Risk Basis

The NRF leverages the findings from the Strategic National Risk Assessment (SNRA) to build and deliver the response core capabilities. The SNRA identifies the threats and hazards that pose the greatest risk to the Nation. These findings affirm the need for an all-hazards, capability-based approach to preparedness to ensure that all types of scenarios are accounted for. The risks and threats identified by SNRA include the following:

- Natural hazards—including hurricanes, earthquakes, wildfires, and floods—present a significant and varied risk across the country.
- A virulent strain of pandemic influenza could kill hundreds of thousands of Americans, affect millions more, and result in considerable economic loss. Additional human and animal infectious diseases, including those previously undiscovered, may also present significant risks.
- Technological and accidental hazards, such as dam failures or chemical substance spills or releases, have the potential to cause extensive fatalities and severe economic impacts, and the likelihood of occurrence may increase due to aging infrastructure.
- Terrorist organizations or affiliates may seek to acquire, build, and use weapons of mass destruction. Conventional terrorist attacks, including those by lone actors employing explosives and armed attacks, present a continued risk to the Nation.
- Cyber attacks can have catastrophic consequences and may also have cascading effects such as power grid or financial system failures.

No single threat or hazard exists in isolation. As an example, a hurricane can lead to flooding, dam failures, and hazardous materials spills. The Goal, therefore, focuses on core capabilities that can be applied to deal with cascading effects. Since many incidents occur with little or no warning, these capabilities must be able to be delivered in a no-notice environment.

In order to establish the basis for these capabilities, planning factors drawn from a number of different scenarios are used to develop the Response FIOP, which supplements the NRF. Refer to the Operational Planning section for additional details on planning assumptions.

Roles and Responsibilities

Effective response depends on integration of the whole community and all partners executing their roles and responsibilities. This section describes those roles and responsibilities and sharpens the focus on identifying who is involved with the Response mission area. It also addresses what the various partners must do to deliver the response core capabilities and to integrate successfully with the Prevention, Protection, Mitigation, and Recovery mission areas.

An effective, unified national response requires layered, mutually supporting capabilities. Individuals and families, communities, the private sector, NGOs, and local, state, tribal, territorial, insular area, and Federal governments should each understand their respective roles and responsibilities and how to complement each other in achieving shared goals. All elements of the whole community play prominent roles in developing the core capabilities needed to respond to incidents. This includes developing plans, conducting assessments and exercises, providing and directing resources and capabilities, and gathering lessons learned. These activities require that all partners understand how they fit within and are supported by the structures described in the NRF.

Emergency management staff in all jurisdictions have a fundamental responsibility to consider the needs of all members of the whole community, including children; individuals with disabilities and others with access and functional needs; those from religious, racial, and ethnically diverse backgrounds; and people with limited English proficiency. The potential contributions of all these individuals toward delivering core capabilities during incident response (e.g., through associations and alliances that serve these populations) should be incorporated into planning efforts.

Staff must also consider those who own or have responsibility for animals both as members of the community who may be affected by incidents and as a potential means of supporting response efforts. This includes those with household pets, service and assistance animals, working dogs, and livestock, as well as those who have responsibility for wildlife, exotic animals, zoo animals, research animals, and animals housed in shelters, rescue organizations, breeding facilities, and sanctuaries.

Individuals, Families, and Households

Although not formally part of emergency management operations, individuals, families, and households play an important role in emergency preparedness and response. By reducing hazards in and around their homes by efforts such as raising utilities above flood level or securing unanchored objects against the threat of high winds, individuals reduce potential emergency response requirements. Individuals, families, and households should also prepare emergency supply kits and emergency plans so they can take care of themselves and their neighbors until assistance arrives. Information on emergency preparedness can be found at many community, state, and Federal emergency management Web sites, such as http://www.ready.gov.

Individuals can also contribute to the preparedness and resilience of their households and communities by volunteering with emergency organizations (e.g., the local chapter of the

American Red Cross, Medical Reserve Corps, or Community Emergency Response Teams [CERTs]) and completing emergency response training courses. Individuals, families, and households should make preparations with family members who have access and functional needs or medical needs. Their plans should also include provisions for their animals, including household pets or service animals. During an actual disaster, emergency, or threat, individuals, households, and families should monitor emergency communications and follow guidance and instructions provided by local authorities.

Communities

Communities are groups that share goals, values, and institutions. They are not always bound by geographic boundaries or political divisions. Instead, they may be faith-based organizations, neighborhood partnerships, advocacy groups, academia, social and community groups, and associations. Communities bring people together in different ways for different reasons, but each provides opportunities for sharing information and promoting collective action. Engaging these groups in preparedness efforts, particularly at the local and state levels, is important to identifying their needs and taking advantage of their potential contributions.

Nongovernmental Organizations

NGOs play vital roles at the local, state, tribal, territorial, insular area government, and national levels in delivering important services, including those associated with the response core capabilities. NGOs include voluntary, racial and ethnic, faith-based, veteran-based, and nonprofit organizations that provide sheltering, emergency food supplies, and other essential support services. NGOs are inherently independent and committed to specific interests and values. These interests and values drive the groups' operational priorities and shape the resources they provide. NGOs bolster government efforts at all levels and often provide specialized services to the whole community, as well as to certain members of the population including children; individuals with disabilities and others with access and functional needs; those from religious, racial, and ethnically diverse backgrounds; and people with limited English proficiency. NGOs are key partners in preparedness activities and response operations. Examples of NGO contributions include:

- Training and managing volunteer resources
- Identifying physically accessible shelter locations and needed supplies to support those displaced by an incident
- Providing emergency commodities and services, such as water, food, shelter, assistance with family reunification, clothing, and supplies for post-emergency cleanup
- Supporting the evacuation, rescue, care, and sheltering of animals displaced by the incident
- Providing search and rescue, transportation, and logistics services and support

- Identifying those whose needs have not been met and helping to provide assistance
- Providing health, medical, mental health, and behavioral health resources
- Assisting, coordinating, and providing disability-related assistance and functional needs support services (FNSS)
- Providing language assistance services to individuals with limited English proficiency.

At the same time that NGOs support response core capabilities, they may also require government assistance. When planning for local community emergency management resources, government organizations should consider the potential need to better enable NGOs to perform their essential response functions. Some NGOs are officially designated as support elements to national response capabilities:

- **The American Red Cross.** The American Red Cross is chartered by Congress to provide relief to survivors of disasters and help people prevent, prepare for, and respond to emergencies. The Red Cross has a legal status of "a federal instrumentality" and maintains a special relationship with the Federal Government. In this capacity, the American Red Cross supports several ESFs and the delivery of multiple core capabilities.
- **National Voluntary Organizations Active in Disaster (VOAD).**[13] National VOAD is the forum where organizations share knowledge and resources throughout the disaster cycle—preparation, response, and recovery—to help disaster survivors and their communities. National VOAD is a consortium of approximately 50 national organizations and 55 territorial and state equivalents.
- **Volunteers and Donations.** Incident response operations frequently exceed the resources of government organizations. Volunteers and donors support response efforts in many ways, and governments at all levels must plan ahead to incorporate volunteers and donated resources into response activities. The goal of volunteer and donations management is to support jurisdictions affected by disasters through close collaboration with the voluntary organizations and agencies. The objective is to manage the influx of volunteers and donations to voluntary agencies and all levels of government before, during, and after an incident. Additional information may be found in the Volunteers and Donations Management Support Annex.

Table 1. Private Sector Roles

Category	Role in This Category
Affected Organization/Component of the Nation's Economy	Private sector organizations may be affected by direct or indirect consequences of an incident. Such organizations include entities that are significant to local, regional, and national economic recovery from an incident. Examples include major employers and suppliers of key commodities or services. As key elements of the national economy, it is important for private sector organizations of all types and sizes to take every precaution necessary to boost resilience, the better to stay in business or resume normal operations quickly.

Table 1. (Continued)

Category	Role in This Category
Affected Infrastructure	Critical infrastructure—such as privately owned transportation and transit, telecommunications, utilities, financial institutions, hospitals, and other health regulated facilities—should have effective business continuity plans.[14]
Regulated and/or Responsible Party	Owners/operators of certain regulated facilities or hazardous operations may be legally responsible for preparing for and preventing incidents and responding when an incident occurs. For example, Federal regulations require owners/operators of nuclear power plants to maintain emergency plans and to perform assessments, notifications, and training for incident response.
Response Resource	Private sector entities provide response resources (donated or compensated) during an incident—including specialized teams, essential services, equipment, and advanced technologies—through local public-private emergency plans or mutual aid and assistance agreements or in response to requests from government and nongovernmental-volunteer initiatives.
Partner With Federal/State/Local Emergency Organizations	Private sector entities may serve as partners in state and local emergency preparedness and response organizations and activities and with Federal sector-specific agencies.
Components of the Nation's Economy	As key elements of the national economy, private sector resilience and continuity of operations planning, as well as recovery and restoration from incidents, represent essential homeland security activities.

Private Sector Entities

Private sector organizations contribute to response efforts through partnerships with each level of government. They play key roles before, during, and after incidents. Private sector entities include large, medium, and small businesses; commerce, private cultural and educational institutions; and industry, as well as public/private partnerships that have been established specifically for emergency management purposes. Private sector organizations may play any of the roles described in Table 1 and may play multiple roles simultaneously. During an incident, key private sector partners should have a direct link to emergency managers and, in some cases, be involved in the decision making process. Strong integration into response efforts can offer many benefits to both the public and private sectors.

A fundamental responsibility of private sector organizations is to provide for the welfare of their employees in the workplace. In addition, some businesses play an essential role in protecting critical infrastructure systems and implementing plans for the rapid reestablishment of normal commercial activities and critical infrastructure operations following a disruption. In many cases, private sector organizations have immediate access to commodities and services that can support incident response, making them key potential contributors of resources necessary to deliver the core capabilities. How the private sector participates in response activities varies based on the type of organization and the nature of the incident.

Examples of key private sector activities include:

- Addressing the response needs of employees, infrastructure, and facilities
- Protecting information and maintaining the continuity of business operations
- Planning for, responding to, and recovering from incidents that impact their own infrastructure and facilities
- Collaborating with emergency management personnel to determine what assistance may be required and how they can provide needed support
- Contributing to communication and information sharing efforts during incidents
- Planning, training, and exercising their response capabilities
- Providing assistance specified under mutual aid and assistance agreements
- Contributing resources, personnel, and expertise; helping to shape objectives; and receiving information about the status of the community.

Local Governments

The responsibility for responding to natural and manmade incidents that have recognizable geographic boundaries generally begins at the local level with individuals and public officials in the county, parish, city, or town affected by an incident. The following paragraphs describe the responsibilities of specific local officials who have emergency management responsibilities.

Chief Elected or Appointed Official

Jurisdictional chief executives are responsible for the public safety and welfare of the people of their jurisdiction. These officials provide strategic guidance and resources across all five mission areas. Chief elected or appointed officials must have a clear understanding of their emergency management roles and responsibilities and how to apply the response core capabilities as they may need to make decisions regarding resources and operations during an incident. Lives may depend on their decisions. Elected and appointed officials also routinely shape or modify laws, policies, and budgets to aid preparedness efforts and improve emergency management and response capabilities. The local chief executive's response duties may include:

- Obtaining assistance from other governmental agencies
- Providing direction for response activities
- Ensuring appropriate information is provided to the public.

Emergency Manager

The jurisdiction's emergency manager oversees the day-to-day emergency management programs and activities. The emergency manager works with chief elected and appointed officials to establish unified objectives regarding the jurisdiction's emergency plans and activities. This role entails coordinating and integrating all elements of the community. The emergency manager coordinates the local emergency management program. This includes assessing the capacity and readiness to deliver the capabilities most likely required during an incident and identifying and correcting any shortfalls. The local emergency manager's duties often include:

- Advising elected and appointed officials during a response
- Conducting response operations in accordance with the NIMS
- Coordinating the functions of local agencies
- Coordinating the development of plans and working cooperatively with other local agencies, community organizations, private sector entities, and NGOs
- Developing and maintaining mutual aid and assistance agreements
- Coordinating resource requests during an incident through the management of an emergency operations center
- Coordinating damage assessments during an incident
- Advising and informing local officials and the public about emergency management activities during an incident
- Developing and executing accessible public awareness and education programs
- Conducting exercises to test plans and systems and obtain lessons learned
- Coordinating integration of the rights of individuals with disabilities, individuals from racially and ethnically diverse backgrounds, and others with access and functional needs into emergency planning and response.

Department and Agency Heads

Department and agency heads collaborate with the emergency manager during the development of local emergency plans and provide key response resources. Participation in the planning process helps to ensure that specific capabilities are integrated into a workable plan to safeguard the community. These department and agency heads and their staffs develop, plan, and train on internal policies and procedures to meet response needs safely. They also participate in interagency training and exercises to develop and maintain necessary capabilities.

State, Tribal, Territorial, and Insular Area Governments

State, tribal, territorial, and insular area governments are responsible for the health and welfare of their residents, communities, lands, and cultural heritage.

States

State governments[15] supplement local efforts before, during, and after incidents by applying in-state resources first. If a state anticipates that its resources may be exceeded, the governor[16] may request assistance from other states or the Federal Government.

Federal assistance may be available to the states under the Stafford Act and other Federal authorities. Under some Federal laws, Federal response actions may be taken without a request from the state. For example, when notified of an oil discharge or chemical release, the Environmental Protection Agency (EPA) and U.S. Coast Guard (USCG) are required to evaluate the need for Federal response and may take action without waiting for a request from state or local officials. Federal financial assistance may also be available to supplement non-

Stafford Act incidents and for disability-related access and functional needs equipment. The following paragraphs describe some of the relevant roles and responsibilities of key officials.

Governor

The public safety and welfare of a state's residents are the fundamental responsibilities of every governor. The governor coordinates state resources and provides the strategic guidance for response to all types of incidents. This includes supporting local governments as needed and coordinating assistance with other states and the Federal Government. A governor also:

- In accordance with state law, may make, amend, or suspend certain orders or regulations associated with response
- Communicates to the public, in an accessible manner (e.g., effective communications to address all members of the whole community), and helps people, businesses, and organizations cope with the consequences of any type of incident
- Coordinates with tribal governments within the state
- Commands the state military forces (National Guard personnel not in Federal service and state militias)
- Coordinates assistance from other states through interstate mutual aid and assistance agreements, such as the Emergency Management Assistance Compact (EMAC)[17]
- Requests Federal assistance including, if appropriate, a Stafford Act declaration of an emergency or major disaster.

State Homeland Security Advisor

Many states have designated homeland security advisors who serve as counsel to the governor on homeland security issues and may serve as a liaison between the governor's office, the state homeland security structure, and other organizations both inside and outside of the state. The advisor often chairs a committee composed of representatives of relevant state agencies, including public safety, the National Guard, emergency management, public health, environment, agriculture, and others charged with developing prevention, protection, mitigation, response, and recovery strategies.

State Emergency Management Agency Director

All states have laws mandating the establishment of a state emergency management agency, as well as the emergency plans coordinated by that agency. The director of the state emergency management agency is responsible for ensuring that the state is prepared to deal with large-scale emergencies and coordinating the statewide response to any such incident. This includes supporting local and tribal governments as needed, coordinating assistance with other states and the Federal Government, and, in some cases, with NGOs and private sector organizations. The state emergency management agency may dispatch personnel to assist in the response and recovery effort.

National Guard

The National Guard is an important state and Federal resource available for planning, preparing, and responding to natural or manmade incidents. National Guard members have expertise in critical areas, such as emergency medical response; communications; logistics;

search and rescue; civil engineering; chemical, biological, radiological, and nuclear response and planning; and decontamination.[18]

The governor may activate elements of the National Guard to support state domestic civil support functions and activities. The state adjutant general may assign members of the Guard to assist with state, regional, and Federal civil support plans.

Other State Departments and Agencies

State department and agency heads and their staffs develop, plan, and train on internal policies and procedures to meet response and recovery needs. They also participate in interagency training and exercises to develop and maintain the necessary capabilities. They are vital to the state's overall emergency management program, as they bring expertise spanning various response functions and serve as core members of the state emergency operations center (EOC) and incident command posts (ICP). Many of them have direct experience in providing accessible and vital services to the whole community during response operations. State departments and agencies typically work in close coordination with their Federal counterpart agencies during joint state and Federal responses, and under some Federal laws, they may request assistance from these Federal partners.

Tribes

The United States has a trust relationship with federally-recognized Indian tribes and recognizes their right to self-government. Tribal governments are responsible for coordinating resources to address actual or potential incidents. When tribal response resources are inadequate, tribal leaders may seek assistance from states or the Federal Government. For certain types of Federal assistance, tribal governments work with the state in which they are located. For other types of Federal assistance, as sovereign entities, tribal governments can elect to work directly with the Federal Government.

Tribes are encouraged to build relationships with local jurisdictions and their states as they may have resources most readily available. The NRF's Tribal Coordination Support Annex outlines processes and mechanisms that tribal governments may use to request direct Federal assistance during an incident regardless of whether or not the incident involves a Stafford Act declaration.

Territories/Insular Areas

Territorial and insular area governments are responsible for coordinating resources to address actual or potential incidents. Due to their remote locations, territories and insular area governments often face unique challenges in receiving assistance from outside the jurisdiction quickly and often request assistance from neighboring islands, other nearby countries, states, the private sector or NGO resources, or the Federal Government. Federal assistance is delivered in accordance with pertinent Federal authorities (e.g., the Stafford Act or through other authorities of Federal departments or agencies).

Tribal/Territorial/Insular Area Leader

The tribal/territorial/insular area government leader is responsible for the public safety and welfare of the people of his/her jurisdiction. As authorized by the tribal, territorial, or insular area government, the leader:

Coordinates resources needed to respond to incidents of all types

- In accordance with the law, may make, amend, or suspend certain orders or regulations associated with the response
- Communicates with the public in an accessible manner and helps people, businesses, and organizations cope with the consequences of any type of incident
- Commands the territory's military forces
- Negotiates mutual aid and assistance agreements with other tribes, territories, insular area governments, states, or local jurisdictions
- Can request Federal assistance under the Stafford Act.

Federal Government

The President leads the Federal Government response effort to ensure that the necessary resources are applied quickly and efficiently to large-scale and catastrophic incidents. The Federal Government maintains a wide range of capabilities and resources that may be required to deal with domestic incidents in a way that ensures the protection of privacy, civil rights, and civil liberties. To be successful, any approach to the delivery of Response capabilities will require an all-of-nation approach. All Federal departments and agencies must cooperate with one another, and with local, state, tribal, territorial, and insular area governments, community members, and the private sector to the maximum extent possible. Although Federal disaster assistance is often considered synonymous with Presidential declarations under the Stafford Act, Federal assistance can actually be provided to state and local jurisdictions, as well as to other Federal departments and agencies, through a number of different mechanisms and authorities.

For incidents in which Federal assistance is provided under the Stafford Act, the Federal Emergency Management Agency (FEMA) coordinates the assistance. For non-Stafford Act incidents, Federal response or assistance may be led or coordinated by various Federal departments and agencies consistent with their authorities.

For incidents on Federal property (e.g., National Parks, military bases) or where the Federal Government has primary jurisdiction, Federal departments or agencies may be the first responders and coordinators of Federal, state, and local activities.

Coordination of Federal Response and Assistance

The President leads the Federal Government response effort to ensure that the necessary resources are applied quickly and efficiently to large-scale and catastrophic incidents. When coordination of Federal response activities is required, it is implemented through the Secretary of Homeland Security, pursuant to Presidential directive.[19] Other Federal departments and agencies carry out their response authorities and responsibilities within this overarching construct.

Secretary of Homeland Security

Pursuant to Presidential directive, the Secretary of Homeland Security is the principal Federal official for domestic incident management. The Secretary coordinates preparedness activities within the United States to respond to and recover from terrorist attacks, major disasters, and other emergencies. As part of these responsibilities, the Secretary coordinates with Federal entities to provide for Federal unity of effort for domestic incident management. The Secretary's responsibilities also include management of the broad "emergency management" and "response" authorities of FEMA and other DHS components.

As part of these responsibilities, the Secretary of Homeland Security also provides the Executive Branch with an overall architecture for domestic incident management and coordinates the Federal response, as required. The Secretary of Homeland Security may monitor activities and activate specific response mechanisms to support other Federal departments and agencies without assuming the overall coordination of the Federal response during incidents that do not require the Secretary to coordinate the response or do not result in a Stafford Act declaration.

Unity of effort differs from unity of command. Various Federal departments and agencies may have statutory responsibilities and lead roles based upon the unique circumstances of the incident. Unity of effort provides coordination through cooperation and common interests and does not interfere with Federal departments' and agencies' supervisory, command, or statutory authorities. The Secretary ensures that overall Federal actions are unified, complete, and synchronized to prevent unfilled gaps or seams in the Federal Government's overarching effort. This coordinated approach ensures that the Federal actions undertaken by DHS and other departments and agencies are harmonized and mutually supportive. The Secretary executes these coordination responsibilities, in part, by engaging directly with the President and relevant Cabinet, department, agency, and DHS component heads as is necessary to ensure a focused, efficient, and unified Federal preparedness posture. All Federal departments and agencies, in turn, cooperate with the Secretary in executing domestic incident management duties.

DHS component heads may have lead response roles or other significant roles depending on the type and severity of the incident. For example, the U.S. Secret Service is the lead agency for security design, planning, and implementation of National Special Security Events (NSSEs) while the Assistant Secretary for Cyber security and Communications coordinates the response to significant cyber incidents.

Other Federal departments and agencies may have a lead or support role in operations coordination. When the Secretary of Homeland Security is not exercising delegated response coordination responsibilities, other Federal departments and agencies may coordinate Federal operations under their own statutory authorities, or as designated by the President, and may activate response structures applicable to those authorities. The head of the department or agency may also request the Secretary to activate other NRF structures and elements to provide additional assistance, while still retaining leadership for the response. For all incidents, Federal department and agency heads serve as advisors to the Executive Branch for their areas of responsibility. Nothing in the NRF precludes any Federal department or agency from executing their authorities.

Several Federal departments and agencies have authorities to respond to and declare specific types of disasters or emergencies apart from the Stafford Act. These authorities may be exercised independently of, concurrently with, or become part of a Federal response coordinated by the Secretary of Homeland Security, pursuant to Presidential directive. Federal departments and agencies carry out their response authorities and responsibilities within the NRF's overarching construct or under supplementary or complementary operational plans. Table 2 provides examples of scenarios in which specific Federal departments and agencies have the responsibility for coordinating response activities. This is not an all-inclusive list. Refer to the NRF Incident Annexes for more details.

Table 2. Examples of Other Federal Department and Agency Authorities

Scenario	Department/Agency	Authorities
Agricultural and Food Incident	Department of Agriculture (USDA)	The Secretary of Agriculture has the authority to declare an **extraordinary emergency** and take action due to the presence of a pest or disease of livestock that threatens livestock in the United States. (7 U.S. Code § 8306 [2007]). The Secretary of Agriculture also has the authority to declare an **extraordinary emergency** and take action due to the presence of a plant pest or noxious weed whose presence threatens plants or plant products of the United States. (7 U.S. Code § 7715 [2007]).
Public Health Emergency[20]	Department of Health and Human Services	The Secretary of the Department of Health and Human Services has the authority to take actions to protect the public health and welfare, declare a **public health emergency** and to prepare for and respond to public health emergencies. (Public Health Service Act, 42 U.S. Code §§ 201 *et seq.*).
Oil and Hazardous Materials Spills	EPA or USCG	The EPA and USCG have the authority to take actions to respond to oil discharges and releases of hazardous substances, pollutants, and contaminants, including leading the response. (42 U.S. Code § 9601, et seq., 33 U.S. Code § 1251 et seq.) The EPA Administrator and Commandant of the USCG[21] may also classify an oil discharge as a **Spill of National Significance** and designate senior officials to participate in the response. (40 CFR § 300.323).[22]

NOTE: These authorities may be exercised independently of, concurrently with, or become part of a Federal response coordinated by the Secretary of Homeland Security pursuant to Presidential directive.

When a Federal department, agency, or component of DHS has responsibility for directing or managing a major aspect of a response coordinated by the Secretary of Homeland Security, that organization is part of the national leadership for the incident and is represented in field, regional, and headquarters unified command and coordination organizations.

FEMA Administrator

The Administrator is the principal advisor to the President, the Secretary of Homeland Security, and the Homeland Security Council regarding emergency management. The FEMA

Administrator's duties include assisting the President, through the Secretary, in carrying out the Stafford Act, operation of the National Response Coordination Center (NRCC), the effective support of all ESFs, and more generally, preparation for, protection against, response to, and recovery from all-hazards incidents. Reporting to the Secretary of Homeland Security, the FEMA Administrator is also responsible for managing the core DHS grant programs supporting homeland security activities.[23]

Attorney General

Like other Executive Branch departments and agencies, the Department of Justice and the Federal Bureau of Investigation (FBI) will endeavor to coordinate their activities with other members of the law enforcement community, and with members of the Intelligence Community, to achieve maximum cooperation consistent with the law and operational necessity.

The Attorney General has lead responsibility for criminal investigations of terrorist acts or terrorist threats, where such acts are within the Federal criminal jurisdiction of the United States. Generally acting through the FBI Director, the Attorney General, in cooperation with Federal departments and agencies engaged in activities to protect our national security, shall also coordinate the activities of the other members of the law enforcement community to detect, prevent, preempt, and disrupt terrorist attacks against the United States.

In addition, the Attorney General approves requests submitted by state governors pursuant to the Emergency Federal Law Enforcement Assistance Act for personnel and other Federal law enforcement support during incidents. The Attorney General also enforces Federal civil rights laws, such as the Americans with Disabilities Act of 1990 and the Civil Rights Act of 1964. Further information on the Attorney General's role is provided in the National Prevention Framework and Prevention FIOP.

Secretary of Defense

Because of the Department of Defense's (DOD's) critical role in national defense, its resources are committed only after approval by the Secretary of Defense or at the direction of the President. Many DOD officials are authorized to respond to save lives, protect property, and mitigate human suffering under imminently serious conditions, as well as to provide support under their separate established authorities, as appropriate.[24] When DOD resources are authorized to support civil authorities, command of those forces remains with the Secretary of Defense. DOD elements in the incident area of operations and the National Guard forces under the command of a governor coordinate closely with response organizations at all levels.

Secretary of State

A domestic incident may have international and diplomatic implications that call for coordination and consultation with foreign governments and international organizations. The Secretary of State is responsible for all communication and coordination between the U.S. Government and other nations regarding the response to a domestic crisis. The Department of State also coordinates international offers of assistance and formally accepts or declines these offers on behalf of the U.S. Government based on needs conveyed by Federal departments and agencies as stated in the International Coordination Support Annex. Some types of international assistance are pre-identified, and bilateral agreements are already established.

For example, the USDA/Forest Service and Department of the Interior have joint bilateral agreements with several countries for wild land firefighting support.

Director of National Intelligence

The Director of National Intelligence serves as the head of the Intelligence Community, acts as the principal advisor to the President for intelligence matters relating to national security, and oversees and directs implementation of the National Intelligence Program. The Intelligence Community, comprising 17 elements across the Federal Government, functions consistent with law, Executive Order, regulations, and policy to support the national security-related missions of the U.S. Government. It provides a range of analytic products, including those that assess threats to the homeland and inform planning, capability development, and operational activities of homeland security enterprise partners and stakeholders. In addition to intelligence community elements with specific homeland security missions, The Office of the Director of National Intelligence maintains a number of mission and support centers that provide unique capabilities for homeland security partners.

Other Federal Department and Agency Heads

Various Federal departments or agencies play primary, coordinating, or support roles in delivering response core capabilities. They may also have responsibilities and authorities to respond independent of any Stafford Act declaration as indicated above. Additional information regarding Federal department and agency roles in delivering core capabilities may be found in the Coordinating Structures and Integration section and in the various annexes to this Framework.

CORE CAPABILITIES

Once an incident occurs, efforts focus on saving lives, protecting property and the environment, and preserving the social, economic, cultural, and political structure of the jurisdiction. Depending on the size, scope, and magnitude of an incident, local, state, tribal, territorial, and insular area governments, and, in some cases, the Federal Government, may be called to action. The response core capabilities are a list of the activities that generally must be accomplished in incident response regardless of which levels of government are involved.

This list was developed based on the results of the SNRA which identified a variety of threats and hazards that would likely stress the Nation's response capabilities. Planners for each mission area—Prevention, Protection, Mitigation, Response, and Recovery—identified functions that would be required to deal with these threats and hazards, and these are the core capabilities. The core capabilities are distinct critical elements necessary to achieve the Goal. They provide a common vocabulary describing the significant functions that must be developed and executed across the whole community to ensure national preparedness.

This section addresses the core capabilities for the Response mission area and the actions required to build and deliver these capabilities.

Context of the Response Mission Area

By engaging the whole community to build and deliver the response core capabilities, the Nation is better prepared to respond to any threat or hazard, assist in restoring basic services and community functionality, and support the transition to recovery. The Response mission area includes 14 core capabilities—11 that apply to response and three that are common to all five mission areas. The Goal assigned specific objectives and performance thresholds for each capability from which metrics will ultimately be identified to track the Nation's progress toward achieving these objectives. Table 3 provides a summary of each response core capability and the critical tasks to achieve its objective.

Table 3. Overview of Response Core Capabilities in the National Preparedness Goal

Core Capabilities and Critical Tasks	
1. Planning *(Cross-cutting with all mission areas)*	**Objective:** Conduct a systematic process engaging the whole community, as appropriate, in the development of executable strategic, operational, and/or community-based approaches to meet defined objectives.
Critical Tasks: • Develop operational plans at the Federal level and in the states and territories that adequately identify critical objectives based on the planning requirements, provide a complete and integrated picture of the sequence and scope of the tasks to achieve the objectives, and are implementable within the time frame contemplated in the plan using available resources.	
2. Public Information and Warning *(Cross-cutting with all mission areas)*	**Objective:** Deliver coordinated, prompt, reliable, and actionable information to the whole community through the use of clear, consistent, accessible, and culturally and linguistically appropriate methods to effectively relay information regarding any threat or hazard and, as appropriate, the actions being taken and the assistance being made available.
Critical Tasks: • Inform all affected segments of society by all means necessary, including accessible tools, of critical lifesaving and life-sustaining information to expedite the delivery of emergency services and aid the public in taking protective actions. • Deliver credible messages to inform ongoing emergency services and the public about protective measures and other life-sustaining actions and facilitate the transition to recovery.	
3. Operational Coordination *(Cross-cutting with all mission areas)*	**Objective:** Establish and maintain a unified and coordinated operational structure and process that appropriately integrates all critical stakeholders and supports the execution of core capabilities.
Critical Tasks: • Mobilize all critical resources and establish command, control, and coordination structures within the affected community and other coordinating bodies in surrounding communities and across the Nation and maintain, as needed, throughout the duration of an incident. • Enhance and maintain NIMS-compliant command, control, and coordination structures to meet basic human needs, stabilize the incident, and transition to recovery.	

Core Capabilities and Critical Tasks	
4. Critical Transportation	**Objective:** Provide transportation (including infrastructure access and accessible transportation services) for response priority objectives, including the evacuation of people and animals, and the delivery of vital response personnel, equipment, and services to the affected areas.
Critical Tasks: • Establish physical access through appropriate transportation corridors and deliver required resources to save lives and to meet the needs of disaster survivors. • Ensure basic human needs are met, stabilize the incident, transition into recovery for an affected area, and restore basic services and community functionality.	
5. Environmental Response/Health and Safety	Objective: Ensure the availability of guidance and resources to address all hazards, including hazardous materials, acts of terrorism, and natural disasters, in support of the responder operations and the affected communities.
Critical Tasks: • Conduct health and safety hazard assessments and disseminate guidance and resources, to include deploying hazardous materials teams, to support environmental health and safety actions for response personnel and the affected population. • Assess, monitor, perform cleanup actions, and provide resources to meet resource requirements and to transition from sustained response to short-term recovery.	
6. Fatality Management Services	**Objective:** Provide fatality management services, including body recovery and victim identification, working with state and local authorities to provide temporary mortuary solutions, sharing information with Mass Care Services for the purpose of reunifying family members and caregivers with missing persons/remains, and providing counseling to the bereaved.
Critical Tasks: • Establish and maintain operations to recover a significant number of fatalities over a geographically dispersed area.	
7. Infrastructure Systems *(Cross-cutting with Recovery mission area)*	**Objective:** Stabilize critical infrastructure functions, minimize health and safety threats, and efficiently restore and revitalize systems and services to support a viable, resilient community.
Critical Tasks: • Decrease and stabilize immediate infrastructure threats to the affected population, to include survivors in the heavily-damaged zone, nearby communities that may be affected by cascading effects, and mass care support facilities and evacuation processing centers with a focus on life-sustainment and congregate care services. • Re-establish critical infrastructure within the affected areas to support ongoing emergency response operations, life sustainment, community functionality, and a transition to recovery.	
8. Mass Care Services	**Objective:** Provide life-sustaining services to the affected population with a focus on hydration, feeding, and sheltering to those with the most need, as well as support for reunifying families.

Table 3. (Continued)

Core Capabilities and Critical Tasks	
Critical Tasks: • Move and deliver resources and capabilities to meet the needs of disaster survivors, including individuals with access and functional needs. • Establish, staff, and equip emergency shelters and other temporary housing options ensuring that shelters and temporary housing units are physically accessible for individuals with disabilities and others with access and functional needs. • Move from congregate care to non-congregate care alternatives, and provide relocation assistance or interim housing solutions for families unable to return to their pre-disaster homes.	
9. Mass Search and Rescue Operations	Objective: Deliver traditional and atypical search and rescue capabilities, including personnel, services, animals, and assets to survivors in need, with the goal of saving the greatest number of endangered lives in the shortest time possible.
Critical Tasks: • Conduct search and rescue operations to locate and rescue persons in distress, based on the requirements of state and local authorities. • Initiate community-based search and rescue support operations across a wide geographically dispersed area. • Ensure the synchronized deployment of local, regional, national, and international teams to reinforce ongoing search and rescue efforts and transition to recovery.	
10. On-scene Security and Protection	**Objective:** Ensure a safe and secure environment through law enforcement and related security and protection operations for people and communities located within affected areas and for all traditional and atypical response personnel engaged in lifesaving and life-sustaining operations.
Critical Tasks: • Establish a safe and secure environment in an affected area. • Provide and maintain on-scene security and meet the protection needs of the affected population over a geographically dispersed area while eliminating or mitigating the risk of further damage to persons, property, and the environment.	
11. Operational Communications	**Objective:** Ensure the capacity for timely communications in support of security, situational awareness, and operations by any and all means available between affected communities in the impact area and all response forces.
Critical Tasks: • Ensure the capacity to communicate with both the emergency response community and the affected populations and establish interoperable voice and data communications between local, state, tribal, territorial, and Federal first responders. • Re-establish sufficient communications infrastructure within the affected areas to support ongoing life-sustaining activities, provide basic human needs, and transition to recovery.	

Core Capabilities and Critical Tasks	
12. Public and Private Services and Resources	Objective: Provide essential public and private services and resources to the affected population and surrounding communities, to include emergency power to critical facilities, fuel support for emergency responders, and access to community staples (e.g., grocery stores, pharmacies, and banks) and fire and other first response services.
Critical Tasks: • Mobilize and deliver governmental, nongovernmental, and private sector resources within and outside of the affected area to save lives, sustain lives, meet basic human needs, stabilize the incident, and transition to recovery, to include moving and delivering resources and services to meet the needs of disaster survivors. • Enhance public and private resource and services support for an affected area.	
13. Public Health and Medical Services	**Objective:** Provide lifesaving medical treatment via emergency medical services and related operations, and avoid additional disease and injury by providing targeted public health and medical support and products to all people in need within the affected area.
Critical Tasks: • Deliver medical countermeasures to exposed populations. • Complete triage and the initial stabilization of casualties and begin definitive care for those likely to survive their injuries. • Return medical surge resources to pre-incident levels, complete health assessments, and identify recovery processes.	
14. Situational Assessment	Objective: Provide all decision makers with decision-relevant information regarding the nature and extent of the hazard, any cascading effects, and the status of the response.
Critical Tasks: • Deliver information sufficient to inform decision making regarding immediate lifesaving and life-sustaining activities, and engage governmental, private, and civic sector resources within and outside of the affected area to meet basic human needs and stabilize the incident. • Deliver enhanced information to reinforce ongoing lifesaving and life-sustaining activities, and engage governmental, private, and civic sector resources within and outside of the affected area to meet basic human needs, stabilize the incident, and transition to recovery.	

No core capability is the responsibility of any one party or single level of government. Each requires an approach that integrates the abilities of elements in the whole community from the individual through the Federal Government, including traditional and non-traditional partners. The Nation must be prepared to deal not only with the normal type of incidents that communities handle every day, but also with incidents of catastrophic proportions. Most of the resources and functions required at the local level to deliver a given core capability are provided by local government agencies with additional members of the community assisting as needed. Catastrophic incidents require many more response assets and engagement with a broader set of partners.[25] Community involvement is vital to providing additional response support. Local residents may well be the primary source of additional manpower in the first hours and days after a catastrophic incident.

Cross-cutting Response Core Capabilities

Three response core capabilities—Planning, Public Information and Warning, and Operational Coordination—span all five mission areas. These common core capabilities are essential to the success of the other core capabilities. They help establish unity of effort among all those involved in the Response mission area.

- **Planning.** Planning makes it possible to manage the life cycle of a potential crisis, determine capability requirements, and help stakeholders learn their roles. It includes the collection, analysis, and dissemination of risk assessment data and the development of plans, procedures, mutual aid and assistance agreements, strategies, and other arrangements to perform specific missions and tasks. Governments at all levels have a responsibility to develop all-hazards response plans prior to and during an incident. Including a broad range of partners in the planning process helps ensure that the needs and potential contributions of all elements are integrated into workable plans.
- **Public Information and Warning.** For an effective response, jurisdictions must provide accurate and accessible information to decision makers and the public. This includes development of accessible message content, such as incident facts, health risk warnings, pre-incident recommendations, evacuation guidance, and other protective measures. It also includes developing strategies for when, where, how, and by whom information will be delivered and ensuring that all levels of government agree on unified messages. Information must be shared with the public and other members of the response community efficiently, effectively, and in an accessible manner. Effective public information and warning is particularly important in dealing with incidents that start small but may evolve to have greater consequences.
- **Operational Coordination.** For incident response, coordination of operations must occur both among those tasked to deliver the various response core capabilities and with those delivering the core capabilities of other mission areas. This coordination occurs through response structures based on clearly established roles, responsibilities, and reporting protocols. Using NIMS principles, structures, and coordinating processes enhances the efficiency and effectiveness of response. Specific actions to achieve this core capability may include coordinating initial actions, managing ESFs, coordinating requests for additional support, and identifying and integrating resources and capabilities.

Integration among Response Core Capabilities and Mission Areas

Interdependencies exist among many of the core capabilities. For example, organizations involved in providing Mass Care Services often rely on resources and functions from organizations that provide Critical Transportation or Public and Private Services and Resources for commodities distribution; Public Information and Warning for messaging, translators, and interpreters; and Operational Communications for reporting and communication that allows shelters to stay in touch with operations centers.

The core capabilities in various mission areas may also be linked through shared assets and services. For example, the functionality provided by geographic information systems can be applied across multiple response core capabilities, as well as core capabilities in the other four mission areas. Thus synergy among mission area resources and processes is important to maximize capabilities and minimize risk. The overarching nature of functions described in these capabilities frequently involves either support to or cooperation of several incident management partners to ensure the seamless integration and transitions among prevention, protection, mitigation, response, and recovery activities.

Potential points of intersection between the Response mission area and other mission areas include the following:

- **Prevention.** Many of the assets that are used on a day-to-day basis to perform intelligence, law enforcement, homeland security, and homeland defense can be applied to support delivery of response core capabilities such as On-Scene Security and Protection and Public Information and Warning.
- **Protection.** Protection of critical infrastructure systems and implementation of plans for the rapid restoration of commercial activities and critical infrastructure operations are crucial aspects of the Protection mission area. Many of the 18 critical infrastructure sectors[26] within the Protection mission area are also represented in the Response mission area. For example, the Public and Private Services and Resources capability depends on private sector owners and operators of critical infrastructure for achieving the capability's objective.
- **Mitigation.** Achieving the mitigation core capability preliminary targets allows for the incorporation of lessons learned in the analysis and planning processes and makes the response core capabilities more resilient and effective.
- **Recovery.** Even while response activities are underway, recovery operations must begin. The emphasis on response gradually gives way to recovery operations; however, recovery core capabilities may involve some of the same functions as response core capabilities. This includes providing essential public health and safety services, restoring interrupted utility and other essential services, reestablishing transportation routes, providing food and shelter for those displaced by an incident, protecting natural and cultural resources and performing environmental compliance, ensuring equal access, reunifying children who have been displaced from their families/guardians, and reopening schools and child care centers.

These overlapping areas are identified through comprehensive planning with the whole community to ensure that they are properly addressed during response to an incident. Ensuring that IOPs properly account for the integration and transition between mission areas is essential.

Response Actions to Deliver Core Capabilities

This section describes the key tasks each major element of the whole community must accomplish to be prepared to deliver the core capabilities. More detailed concepts of operations for the delivery of the core capabilities are provided in the Response FIOP and operational plans developed by various jurisdictions, the private sector, and NGOs.

Individuals and Households

Many individuals have talents and experience that can be tapped to support core capabilities. Individuals can contribute to the delivery of response core capabilities through community organizations, by participating in community preparedness activities, such as CERT, and by ensuring that they have household/family emergency plans.[27]

Private Sector

Roles and responsibilities of private sector entities are described in the Roles and Responsibilities section. Private sector entities can assist in delivering the response core capabilities by collaborating with emergency management personnel before an incident occurs to determine what assistance may be necessary and how they can support local emergency management during response operations.[28]

Nongovernmental Organizations

NGOs manage volunteers and resources that bolster government efforts to ensure a successful incident response. Collaboration with responders, governments at all levels, and other agencies and organizations helps NGOs to tailor and direct their efforts that are necessary to accomplish and deliver the response core capabilities.

State, Tribal, and Local Actions

Communities apply NIMS principles to integrate response plans and resources across jurisdictions and departments as well as with the private sector and NGOs. Neighboring communities play a key role by providing support through a network of mutual aid and assistance agreements that identify the resources that communities may be able to share during an incident.

The state is the gateway to many government resources that help communities respond. When an incident grows or has the potential to grow beyond the capability of a local jurisdiction and responders cannot meet the needs with mutual aid and assistance resources, local officials contact the state. Upon receiving a request for assistance from a local or tribal government, state officials may:

- Coordinate warnings and public information through the activation of the state's public communications strategy
- Distribute supplies stockpiled to meet the needs of the emergency
- Provide technical assistance and support to meet the response and recovery needs
- Suspend or waive statutes, rules, ordinances, and orders, to the extent permitted by law, to ensure timely performance of response functions
- Implement state volunteer and donations management plans, and coordinate with the private sector and NGOs
- Order or recommend evacuations ensuring the integration and inclusion of the requirements of populations such as: children, individuals with disabilities and others with access and functional needs, those from religious, racial, and ethnically diverse communities, people with limited English proficiency, and owners of animals including household pets and service animals

- Mobilize resources to meet the requirements of individuals with disabilities and others with access and functional needs in compliance with Federal civil rights laws.

If local resources are inadequate, local authorities may seek assistance from the county emergency manager or the state. Under some Federal authorities, local jurisdictions may also seek assistance directly from the Federal Government for non-Stafford Act incidents.

State-to-State Assistance

If additional resources are required, states request assistance from other states through interstate mutual aid and assistance agreements such as EMAC. Administered by the National Emergency Management Association, EMAC is an interstate mutual aid agreement that streamlines the interstate mutual aid and assistance process.

Federal Actions

In certain circumstances, Federal departments and agencies may provide assistance or even lead response efforts consistent with their own authorities.

Federal Response and Assistance under the Stafford Act

When an incident is anticipated to exceed state resources or when the Federal Government has unique capabilities needed by states, the governor may request Federal assistance. In such cases, the affected local jurisdiction and the state, tribal, territorial, insular area, and Federal governments coordinate to provide the necessary assistance. The Federal Government may provide assistance in the form of funding, resources, and services. Federal departments and agencies respect the sovereignty and responsibilities of local, state, tribal, territorial, and insular area governments while rendering assistance that supports the affected local or state governments.

Robert T. Stafford Disaster Relief and Emergency Assistance Act

Local, state, tribal, territorial, and insular area governments do not require Federal assistance to respond to most incidents; however, when an incident is of such severity and magnitude that effective response is beyond the capabilities of the state and local governments, the governor can request Federal assistance under the Stafford Act. In certain circumstances, the President may declare an emergency without a request from a governor when the primary responsibility for response rests with the United States because the emergency involves a subject area for which, under the Constitution or laws of the United States, the United States exercises exclusive or preeminent responsibility and authority.

The Stafford Act authorizes the President to provide financial and other assistance to local, state, tribal, territorial, and insular area governments, certain private nonprofit organizations, and individuals to support response, recovery, and mitigation efforts following a Stafford Act Emergency or Major Disaster Declaration.[29] Most forms of Stafford Act assistance require a state cost share. While Federal assistance under the Stafford Act may only be delivered after a declaration, FEMA may pre-deploy Federal assets when a declaration is likely and imminent. The Stafford Act provides for two types of declarations:

- An **Emergency Declaration** is more limited in scope than a Major Disaster Declaration, provides fewer Federal programs, and is not normally associated with recovery programs. However, the President may issue an Emergency Declaration prior to an actual incident to lessen or avert the threat of a catastrophe. Generally, Federal assistance and funding are provided to meet specific emergency needs or to help prevent a catastrophe from occurring.
- A **Major Disaster Declaration** provides more Federal programs for response and recovery than an Emergency Declaration. Unlike an Emergency Declaration, a Major Disaster Declaration may only be issued after an incident.

Requesting a Stafford Act Declaration

Before requesting a declaration under the Stafford Act, the situation or disaster must be of such severity and magnitude that an effective response is beyond the capabilities of the state and the affected local governments and requires Federal assistance. The governor must take appropriate response action under state law and direct execution of the state's emergency plan. Ordinarily, the governor must ensure certain state and local actions have been taken or initiated, including:

- Surveying the affected areas to determine the extent of private and public damage
- Conducting joint preliminary damage assessments with FEMA officials to estimate the types and extent of Federal disaster assistance required
- Agreeing to provide, without cost to the Federal Government, easements and rights-of-way necessary to accomplish the work and to indemnify the Federal Government against any claims arising from such work when requesting direct Federal assistance
- Agreeing to pay the state's cost share.

The state's request for a Stafford Act declaration, addressed to the President, is submitted through the FEMA Regional Administrator, who evaluates the request and makes a recommendation to the FEMA Administrator. The FEMA Administrator, in coordination with the Secretary of Homeland Security, then makes a recommendation to the President. The governor is notified when the President has acted on the state's request and the appropriate members of Congress and Federal departments and agencies are also notified.

U.S. territories may use the same incident management and response structures and mechanisms as state governments for requesting and receiving Federal assistance. U.S. territories often pose special response challenges. Working in partnership with territorial governments, the processes and structures described in the NRF can be adapted to meet these geographic challenges through preparedness plans and the pre-staging of assets.

Territorial governments may receive federally coordinated response for U.S. possessions, including insular area governments. The freely associated states of the Federated States of Micronesia and the Republic of the Marshall Islands[30] may also receive assistance. Stafford Act assistance is available to Puerto Rico, the U.S. Virgin Islands, Guam, American Samoa, and the Commonwealth of the Northern Mariana Islands, which are included in the definition of "state" in the Stafford Act.

Proactive Response to Catastrophic Incidents

Prior to and during catastrophic incidents, especially those that occur with little or no notice, the Federal Government may mobilize and deploy assets in anticipation of a formal request from the state. Such deployments of significant Federal assets would occur in anticipation of or following catastrophic incidents involving chemical, biological, radiological, nuclear, or high-yield explosive weapons of mass destruction; large-magnitude earthquakes; or other incidents affecting heavily populated areas. Proactive efforts are intended to ensure that Federal resources reach the scene in time to assist in reducing disruption of normal functions of state and local governments and are done in coordination and collaboration with state and local governments, private sector entities, and NGOs when possible.

Federal Response and Assistance Available Without a Stafford Act Declaration

The NRF covers the full range of complex and constantly changing requirements in anticipation of, or in response to, threats or actual incidents. In addition to Stafford Act support, the NRF or other supplementary or complementary operational plans may be applied to respond or provide other forms of support.

Federal Departments and Agencies Acting under Their Own Authorities

Immediate lifesaving assistance to states, as well as other types of assistance, such as wild land firefighting support or response to an agricultural disease or cyber security incident, are performed by Federal departments or agencies under their own authorities and funding or through reciprocal mutual assistance agreements and do not require a Stafford Act declaration. Some Federal departments or agencies conduct or may lead Federal response actions under their own authorities using funding sources other than the President's Disaster Relief Fund. For example, specific trust funds are established under Federal environmental laws to support and fund oil and hazardous substances response operations.

Federal-to-Federal Support

Federal departments and agencies may execute interagency or intra-agency reimbursable agreements in accordance with the Economy Act or other applicable authorities. The Financial Management Support Annex to the NRF contains information about this process. A Federal department or agency responding to an incident under its own authorities may also request support from the Secretary of Homeland Security in obtaining and coordinating additional Federal assistance. The Secretary of Homeland Security may activate one or more ESFs to provide the requested support.

COORDINATING STRUCTURES AND INTEGRATION

Coordinating structures aid preparedness and response at all levels of government and within the private sector, communities, and nongovernmental entities. The structures help organize and measure the whole community's capabilities in order to address the requirements of the Response mission area, facilitate problem solving, improve access to response resources, and foster coordination prior to and following an incident.

Scalable, flexible, and adaptable coordinating structures are essential in aligning the key roles and responsibilities to deliver the Response mission area's core capabilities. The flexibility of such structures helps ensure that communities across the country can organize response efforts to address a variety of risks based on their unique needs, capabilities, demographics, governing structures, and non-traditional partners. The NRF is not based on a one-size-fits-all organizational construct, but instead acknowledges the concept of tiered response, which emphasizes that response to incidents should be handled at the lowest jurisdictional level capable of handling the mission. These structures can be partially or fully implemented in the context of a threat, in anticipation of a significant event, or in response to an incident. Selective implementation allows for a scaled response, delivery of the exact resources that are needed, and a level of coordination appropriate to each incident.

The following section describes the coordinating structures within the Response mission area and explains how they integrate with the coordinating structures that support other mission areas to build preparedness and enhance the Nation's resilience to all types of risks and hazards.

Local Coordinating Structures

Local jurisdictions and states employ a variety of coordinating structures to help identify risks, establish relationships, organize, and build capabilities. Due to the unique partnerships, geographic conditions, threats, and established capabilities each jurisdiction faces, the coordinating structures at these levels vary. Examples of local response coordinating structures include local planning committees, CERTs, and chapters of national-level associations. These structures organize and integrate their capabilities and resources with neighboring jurisdictions, the state, the private sector, and NGOs.

State Coordinating Structures

States also leverage the capabilities and resources of partners across the state when identifying needs and building capabilities. The coordinating structures at the state level also vary depending on factors such as geography, population, industry, and the capabilities of the local jurisdictions within the state. These structures are also designed to leverage appropriate representatives from across the whole community—some of whom may also participate in local or regional coordinating structures. Many states create independent committees or councils focused on specific areas or functions as a sub-set of their emergency management agency. For example, some states have Animal Disaster Planning Advisory Committees that provide important input to statewide response plans on animal issues.

Private Sector Coordinating Structures

Business EOCs, industry trade groups, and private sector information and intelligence centers serve as coordinating structures for the private sector. These organizations, composed of multiple businesses and entities brought together by shared geography or common function (e.g., banking, supply chain management, transportation, venue management), support the collaboration, communication, and sharing of information within the private sector. Such organizations can coordinate with and support NGOs, and in many cases they serve as a conduit to local and state government coordinating structures.

Federal Coordinating Structures

National Security Council

The National Security Council (NSC) is the principal policy body for consideration of national security policy issues requiring Presidential determination. The NSC advises and assists the President in integrating all aspects of national security policy as it affects the United States—domestic, foreign, military, intelligence, and economic (in conjunction with the National Economic Council). Along with its subordinate committees, the NSC is the President's principal means for coordinating Executive Branch departments and agencies in the development and implementation of national security policy.

Emergency Support Functions

The Federal Government and many state governments organize their response resources and capabilities under the ESF construct. ESFs have proven to be an effective way to bundle and manage resources to deliver core capabilities. The Federal ESFs are the primary, but not exclusive, Federal coordinating structures for building, sustaining, and delivering the response core capabilities. The ESFs are vital structures for responding to Stafford Act incidents; however, they may also be used for other incidents. Most Federal ESFs support a number of the response core capabilities. In addition, there are responsibilities and actions associated with Federal ESFs that extend beyond the core capabilities and support other response activities, as well as department and agency responsibilities.

The Federal ESFs bring together the capabilities of Federal departments and agencies and other national-level assets. ESFs are not based on the capabilities of a single department or agency, and the functions for which they are responsible cannot be accomplished by any single department or agency. Instead, Federal ESFs are groups of organizations that work together to deliver core capabilities and support an effective response.

As noted above, many state and local jurisdictions have adopted and tailored the ESF construct. Because state and local jurisdictions establish ESFs based on their specific risks and requirements, there is no mandatory or direct linkage to the Federal ESFs. State and local governments are encouraged to engage members of the whole community as part of whatever coordinating processes they use.

Table 4 summarizes the Federal ESFs and indicates the response core capabilities each ESF most directly supports. All ESFs support the common core capabilities–Planning, Public Information and Warning, and Operational Coordination–and many ESFs support more than those that are listed.

ESF Member Roles and Responsibilities

ESFs are not solely attributed to any one organization, nor are they mechanisms for executing an agency's statutory authorities. Each ESF is composed of a department or agency that has been designated as the ESF coordinator along with a number of primary and support agencies. Primary agencies are designated on the basis of their authorities, resources, and capabilities. Support agencies are assigned based on resources or capabilities in a given functional area. To the extent possible, resources provided by the ESFs are identified consistently with NIMS resource typing categories.

Table 4. Emergency Support Functions and ESF Coordinators

ESF #1—Transportation **ESF Coordinator: Department of Transportation**
Key Response Core Capability: Critical Transportation
Coordinates the support of management of transportation systems and infrastructure, the regulation of transportation, management of the Nation's airspace, and ensuring the safety and security of the national transportation system. Functions include but are not limited to: • Transportation modes management and control • Transportation safety • Stabilization and reestablishment of transportation infrastructure • Movement restrictions • Damage and impact assessment.
ESF #2—Communications **ESF Coordinator: DHS/National Communications System**
Key Response Core Capability: Operational Communications
Coordinates the reestablishment of the critical communications infrastructure, facilitates the stabilization of systems and applications from cyber attacks, and coordinates communications support to response efforts. Functions include but are not limited to: • Coordination with telecommunications and information technology industries • Reestablishment and repair of telecommunications infrastructure • Protection, reestablishment, and sustainment of national cyber and information technology resources • Oversight of communications within the Federal response structures.
ESF #3—Public Works and Engineering **ESF Coordinator: DOD/U.S. Army Corps of Engineers**
Key Response Core Capabilities: Infrastructure Systems, Critical Transportation, Public and Private Services and Resources, Environmental Response/Health and Safety, Fatality Management, Mass Care Services, Mass Search and Rescue Operations
Coordinates the capabilities and resources to facilitate the delivery of services, technical assistance, engineering expertise, construction management, and other support to prepare for, respond to, and/or recover from a disaster or an incident. Functions include but are not limited to: • Infrastructure protection and emergency repair • Critical infrastructure reestablishment • Engineering services and construction management • Emergency contracting support for lifesaving and life-sustaining services.
ESF #4—Firefighting **ESF Coordinator: USDA/U.S. Forest Service and DHS/FEMA/U.S. Fire Administration**
Key Response Core Capabilities: Critical Transportation, Operational Communications, Public and Private Services and Resources, Infrastructure Systems, Mass Care Services, Mass Search and Rescue Operations, On-scene Security and Protection, Public Health and Medical Services

Coordinates the support for the detection and suppression of fires. Functions include but are not limited to: • Support to wild land, rural, and urban firefighting operations.
ESF #5—Information and Planning **ESF Coordinator: DHS/FEMA**
Key Response Core Capabilities: Situational Assessment, Planning, Public Information and Warning
Supports and facilitates multiagency planning and coordination for operations involving incidents requiring Federal coordination. Functions include but are not limited to: • Incident action planning • Information collection, analysis, and dissemination.
ESF #6—Mass Care, Emergency Assistance, Temporary Housing, and Human Services **ESF Coordinator: DHS/FEMA**
Key Response Core Capabilities: Mass Care Services, Public and Private Services and Resources, Public Health and Medical Services, Critical Transportation, Fatality Management Services
Coordinates the delivery of mass care and emergency assistance, including: • Mass care • Emergency assistance • Disaster housing • Human services.
ESF #7—Logistics **ESF Coordinator: General Services Administration and DHS/FEMA**
Key Response Core Capabilities: Public and Private Services and Resources, Mass Care Services, Critical Transportation, Infrastructure Systems, Operational Communications
Coordinates comprehensive incident resource planning, management, and sustainment capability to meet the needs of disaster survivors and responders. Functions include but are not limited to: • Comprehensive, national incident logistics planning, management, and sustainment capability • Resource support (e.g., facility space, office equipment and supplies, contracting services).
ESF #8—Public Health and Medical Services **ESF Coordinator: Department of Health and Human Services**
Key Response Core Capabilities: Public Health and Medical Services, Fatality Management Services, Mass Care Services, Critical Transportation, Public Information and Warning, Environmental Response/Health and Safety, Public and Private Services and Resources
Coordinates the mechanisms for assistance in response to an actual or potential public health and medical disaster or incident. Functions include but are not limited to: • Public health • Medical surge support including patient movement • Behavioral health services • Mass fatality management.
ESF #9—Search and Rescue **ESF Coordinator: DHS/FEMA**
Key Response Core Capability: Mass Search and Rescue Operations

Table 4. (Continued)

Coordinates the rapid deployment of search and rescue resources to provide specialized lifesaving assistance. Functions include but are not limited to: • Structural Collapse (Urban) Search and Rescue • Maritime/Coastal/Waterborne Search and Rescue • Land Search and Rescue.
ESF #10—Oil and Hazardous Materials Response **ESF Coordinator: Environmental Protection Agency**
Key Response Core Capabilities: Environmental Response/Health and Safety, Critical Transportation, Infrastructure Systems, Public Information and Warning
Coordinates support in response to an actual or potential discharge and/or release of oil or hazardous materials. Functions include but are not limited to: • Environmental assessment of the nature and extent of oil and hazardous materials contamination • Environmental decontamination and cleanup.
ESF #11—Agriculture and Natural Resources **ESF Coordinator: Department of Agriculture**
Key Response Core Capabilities: Environmental Response/Health and Safety, Mass Care Services, Public Health and Medical Services, Critical Transportation, Public and Private Services and Resources, Infrastructure Systems
Coordinates a variety of functions designed to protect the Nation's food supply, respond to plant and animal pest and disease outbreaks, and protect natural and cultural resources. Functions include but are not limited to: • Nutrition assistance • Animal and agricultural health issue response • Technical expertise, coordination, and support of animal and agricultural emergency management • Meat, poultry, and processed egg products safety and defense • Natural and cultural resources and historic properties protection.
ESF #12—Energy **ESF Coordinator: Department of Energy**
Key Response Core Capabilities: Infrastructure Systems, Public and Private Services and Resources, Situational Assessment
Facilitates the reestablishment of damaged energy systems and components and provides technical expertise during an incident involving radiological/nuclear materials. Functions include but are not limited to: • Energy infrastructure assessment, repair, and reestablishment • Energy industry utilities coordination • Energy forecast.
ESF #13—Public Safety and Security **ESF Coordinator: Department of Justice/Bureau of Alcohol, Tobacco, Firearms, and Explosives**
Key Response Core Capability: On-scene Security and Protection
Coordinates the integration of public safety and security capabilities and resources to support the full range of incident management activities. Functions include but are not limited to: • Facility and resource security

• Security planning and technical resource assistance • Public safety and security support • Support to access, traffic, and crowd control.
ESF #14—Superseded by National Disaster Recovery Framework
ESF #15—External Affairs **ESF Coordinator: DHS**
Key Response Core Capability: Public Information and Warning
Coordinates the release of accurate, coordinated, timely, and accessible public information to affected audiences, including the government, media, NGOs, and the private sector. Works closely with state and local officials to ensure outreach to the whole community. Functions include, but are not limited to: • Public affairs and the Joint Information Center • Intergovernmental (local, state, tribal, and territorial) affairs • Congressional affairs • Private sector outreach • Community relations.

- **ESF Coordinators**. ESF coordinators oversee the preparedness activities for a particular ESF and coordinate with its primary and support agencies. Responsibilities of the ESF coordinator include:
  - Maintaining contact with ESF primary and support agencies through conference calls, meetings, training activities, and exercises
  - Monitoring the ESF's progress in meeting the targets of the core capabilities it supports
  - Coordinating efforts with corresponding private sector, NGO, and Federal partners
  - Ensuring the ESF is engaged in appropriate planning and preparedness activities.
- **Primary Agencies.** ESF primary agencies have significant authorities, roles, resources, and capabilities for a particular function within an ESF. Primary agencies are responsible for:
  - Orchestrating support within their functional area for the appropriate response core capabilities and other ESF missions
  - Notifying and requesting assistance from support agencies
  - Managing mission assignments (in Stafford Act incidents) and coordinating with support agencies, as well as appropriate state officials, operations centers, and other stakeholders
  - Coordinating resources resulting from mission assignments
  - Working with all types of organizations to maximize the use of all available resources
  - Monitoring progress in achieving core capability targets and other ESF missions, and providing that information as part of situational and periodic readiness or preparedness assessments
  - Planning for incident management, short-term recovery operations, and the transition to long-term recovery
  - Maintaining trained personnel to support interagency emergency response and support teams

    - Identifying new equipment or capabilities required to prevent or respond to new or emerging threats and hazards or to validate and improve capabilities to address changing risks
    - Promoting physical accessibility, programmatic inclusion, and effective communication for the whole community, including individuals with disabilities.
- **Support Agencies.** ESF support agencies have specific capabilities or resources that support primary agencies in executing the mission of the ESF. The activities of support agencies typically include:
    - Participating in planning for incident management, short-term recovery operations, transition to long-term-recovery, and the development of supporting operational plans, standard operating procedures (SOPs), checklists, or other job aids
    - Providing input to periodic readiness assessments
    - Maintaining trained personnel to support interagency emergency response and support teams
    - Identifying new equipment or capabilities required to respond to new or emerging threats and hazards, or to improve the ability to address existing threats
    - Coordinating resources resulting from response mission assignments.
- **Emergency Support Function Leaders Group (ESFLG).** The ESFLG comprises the Federal departments and agencies that are designated as coordinators for ESFs or coordinating agencies for other NRF annexes. FEMA leads the ESFLG and is responsible for calling meetings and other administrative functions. The ESFLG provides a forum for departments and agencies with roles in Federal incident response to jointly address topics such as policies, preparedness, and training.

ESF Activation

Departments and agencies supporting Federal ESFs may be selectively activated by FEMA or as directed by the Secretary of Homeland Security to support response activities for both Stafford Act and non-Stafford Act incidents. Not all incidents requiring Federal support result in the activation of ESFs.

When departments and agencies supporting Federal ESFs are activated, they may assign staff at headquarters, regional, and incident levels. Through the Stafford Act and in accordance with 6 U.S. Code 741 (4) and 753 (c), FEMA may issue mission assignments at all levels to obtain resources and services from Federal departments and agencies across the ESFs.

ESFs are the primary, but not exclusive, response coordinating structures at the Federal level for Stafford Act incidents.

Communities, states, regions, and other Federal departments and agencies may use the ESF construct, or they may employ other coordinating structures or partners appropriate to their location, threats, or authorities. Whatever structures are used, they are encouraged to work closely with Federal ESFs at the incident, regional, or headquarters levels if they are activated.

One example of a unique Federal coordinating structure is described below:

- **Tribal Assistance Coordination Group (TAC-G).** Governments at the Federal, state, and local levels foster effective government-to-government working relationships with tribes to achieve the common goal of responding to disasters impacting tribal lands. The TAC-G comprises multiple Federal organizations that are dedicated to cooperation and collaboration to strengthen emergency management as it relates to the over 560 federally-recognized tribal nations.

Non-Stafford Act Coordinating Structures

Although the Federal ESFs are designed to coordinate Federal response resources for both Stafford Act and non-Stafford Act incidents, the ESFs may not always be the most appropriate response coordinating structures for non-Stafford Act incidents. For incidents in which there is no Stafford declaration, the department or agency with primary legal authority may activate the coordinating structures appropriate to that authority. These structures are generally organized consistently with NIMS concepts and principles. In addition to their own structures, departments or agencies responding under their own legal authorities may request the Secretary of Homeland Security to activate relevant ESFs. Pursuant to Presidential directive, the Secretary of Homeland Security coordinates with the head of the department or agency with primary legal authority but retains the authority to activate ESFs or other coordinating structures, as appropriate.

NRF Support Annexes

The NRF Support Annexes describe other mechanisms by which support is organized among private sector, NGO, and Federal partners. Federal departments and agencies designated as coordinating and cooperating agencies in NRF support annexes conduct a variety of activities to include managing specific functions and missions and providing Federal support within their functional areas. The Support Annexes include:

- Critical Infrastructure
- Financial Management
- International Coordination
- Private Sector Coordination
- Tribal Coordination
- Volunteer and Donations Management
- Worker Safety and Health.

NRF Incident Annexes

NRF Incident Annexes describe coordinating structures, in addition to the ESFs, that may be used to deliver core capabilities and support response missions that are unique to a specific type of incident. Incident annexes also describe specialized response teams and resources, incident-specific roles and responsibilities, and other scenario-specific considerations. NRF Incident Annexes address the following contingencies or hazards:

- Biological Incident
- Catastrophic Incident

- Cyber Incident
- Food and Agriculture Incident
- Mass Evacuation Incident
- Nuclear/Radiological Incident
- Terrorism Incident Law Enforcement and Investigation.

Operations Coordination

Response operations involve multiple partners and stakeholders. Operations coordination occurs at all government levels and consists of actions and activities that enable decision makers to determine appropriate courses of action and provide oversight for complex homeland security operations to achieve unity of effort and effective outcomes.

Local Response Operational Structures

Emergency responders at all levels of government use ICS command and coordinating structures to manage response operations (see Figure 2). ICS is a management system designed to integrate facilities, equipment, personnel, procedures, and communications within a common organizational structure. At the local level, coordinating structures are usually composed of entities within a specific functional area such as public works, law enforcement, emergency medical services, and fire departments. Integration among these structures occurs at an ICP, which provides on-scene incident command and management. ICS is widely used by all levels of government, as well as by private sector organizations and NGOs to organize field-level operations for a broad spectrum of incidents. Typically, the incident response is structured to facilitate activities in five areas: command, operations, planning, logistics, and finance/administration. Emergency personnel may also use the Multiagency Coordination System (MACS). The primary function of MACS, as defined in NIMS, is to coordinate activities above the incident level and to prioritize competing demands for incident resources. MACS consist of personnel, procedures, protocols, facilities, business practices, and communications integrated into a common system. MACS elements at the local level include EOCs and coordination centers.

If the local incident commander determines that additional resources or capabilities are needed, he or she contacts the local EOC and relays requirements to the local emergency manager. Local EOCs are the physical locations where multiagency coordination typically occurs and where a variety of local coordinating structures come together to solve problems. EOCs help form a common operating picture of the incident, relieve on-scene command of the burden of external coordination, and secure additional resources to help meet response requirements.

EOCs at all levels of government may also encourage participation by the private sector, NGOs, academia, associations, racial and ethnic organizations, and access and functional needs subject matter experts. These members of the whole community, in turn, often maintain their own structures, such as nongovernmental or private sector EOCs.

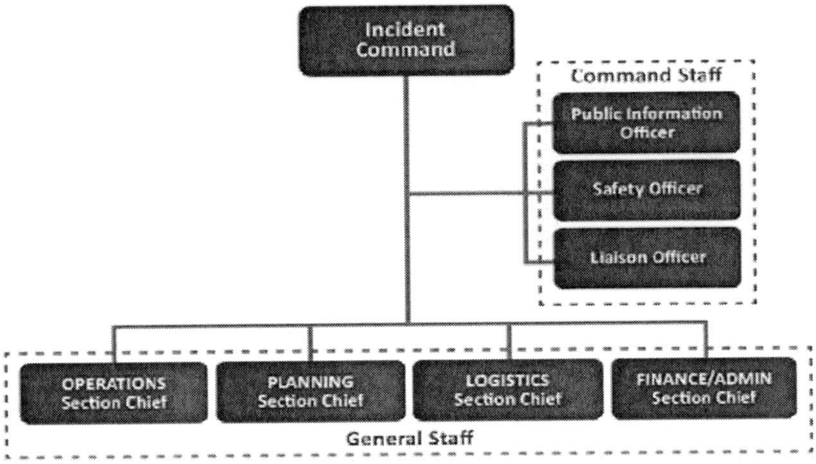

Figure 2. Incident Command Structure.

State Response Operational Structures

The local incident command structure directs on-scene incident management activities and maintains command and control of on-scene incident operations. State EOCs are activated as necessary to support local EOCs and to ensure that responders have the resources they need to conduct response activities. This is achieved through integration of state-level coordinating structures working with local coordinating structures or the local incident command structure.

State Emergency Operations Center

State EOCs are the physical location where multiagency coordination occurs through state-level coordinating structures. Every state maintains an EOC to manage incidents requiring state-level assistance (see Figure 3). Some states have additional EOCs for coordinating information and resources within a region or area. State EOCs are typically organized by a combination of ESFs or other coordinating structures aligned to disciplines or capabilities. Many states involve their tribal counterparts within the EOC to ensure that tribal coordinating structures are integrated into the delivery of capabilities and that tribal needs are addressed.

Federal Response Operational Structures

When an incident occurs that exceeds, or is anticipated to exceed, local or state resources—or when an incident is managed by Federal departments or agencies acting under their own authorities—the Federal Government may use the management structures described within the NRF. Additionally, the Federal Government may use supplementary or complementary plans to involve all necessary department and agency resources to organize the Federal response and ensure coordination among all response partners. All Federal departments and agencies may play significant roles in response activities depending on the nature and size of an incident. Many of the arrangements by which departments and agencies participate are defined in the ESF Annexes, coordinated through pre-scripted mission assignments in a Stafford Act response, formalized in interagency agreements, or described in NRF supplementary plans.

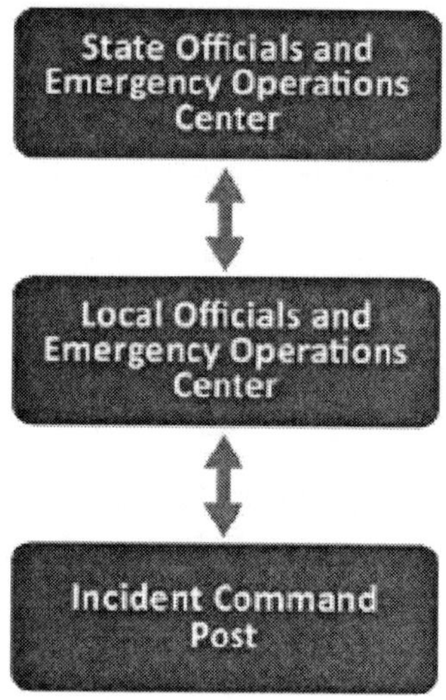

Figure 3. State and Local Response Structure.

The following sections describe Federal support operations at the incident, regional, and headquarters levels.

Federal Incident-level Operations

To help deliver Federal support or response at the incident level, coordinating structures are aligned to incident-level structures. The following section describes the Federal coordinating structures typically associated with Stafford Act incidents. These structures may also be used for Federal-to-Federal support or other non-Stafford Act threats or incidents such as an NSSE.

Unified Coordination

Unified Coordination is the term used to describe the primary state/Federal incident management activities conducted at the incident level. Unified Coordination is typically directed from a Joint Field Office (JFO), a temporary Federal facility that provides a central location for coordination of response efforts by the private sector, NGOs, and all levels of government. Unified Coordination is organized, staffed, and managed in a manner consistent with NIMS principles using the NIMS/ICS structure. The Unified Coordination Group (UCG) comprises senior leaders representing Federal and state interests and, in certain circumstances, tribal governments, local jurisdictions, and the private sector. UCG members must have significant jurisdictional responsibility and authority. The composition of the UCG varies from incident to incident depending on the scope and nature of the disaster. The UCG leads the unified coordination staff. Personnel from state and Federal departments and agencies, other jurisdictional entities, the private sector, and NGOs may be assigned to the unified coordination staff at various incident facilities (e.g., JFO, staging areas, and other field offices). The UCG determines staffing of the unified coordination staff based on incident requirements.

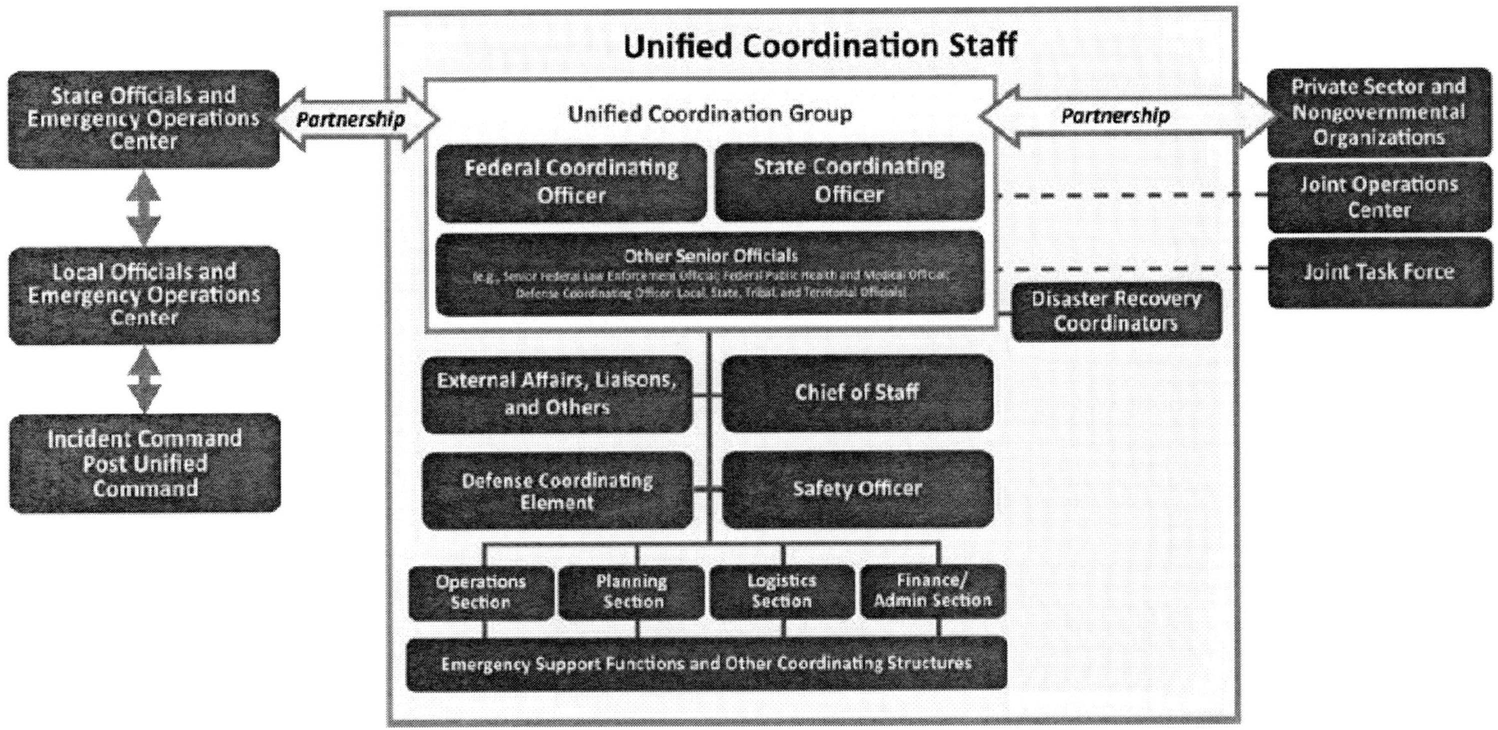

Figure 4. Unified Coordination.

Although Unified Coordination is based on the ICS structure, it does not manage on-scene operations. Instead, it focuses on providing support to on-scene response efforts and conducting broader support operations that may extend beyond the incident site. Unified Coordination must include robust operations, planning, public information, and logistics capabilities that integrate local, state, and Federal—as well as tribal, territorial, and insular area governments—personnel when appropriate, so that all levels of government work together to achieve unity of effort.

When incidents affect multiple localities and states or the entire Nation, multiple UCGs with associated unified coordination staff may be established. In these situations, coordination occurs according to the principles of area command as described in NIMS.

As the primary field entity for Federal response, Unified Coordination integrates diverse Federal authorities and capabilities and coordinates Federal response and recovery operations. Figure 4 depicts a Unified Coordination organization that might be assembled to deal with a major incident, such as a terrorist attack, that includes a law enforcement dimension. Federal agencies that conduct on-scene, tactical level activities under the Stafford Act may also establish incident and area command structures, generally in conjunction with their counterpart local, state, tribal, and/or insular area agencies, to manage that work.

Federal Incident-level Operations for Non-Stafford Act Incidents

For non-Stafford Act incidents, the department or agency with primary legal jurisdiction activates the response structures appropriate to its authorities; these structures are generally organized based on NIMS concepts and principles. When coordinating pursuant to Presidential directive, the Secretary of Homeland Security coordinates with the head of the department or agency with primary legal jurisdiction but retains the authority to activate the additional response structures the Secretary determines appropriate.

Federal Incident Command/Area Command in Non-Stafford Act Incidents

In non-Stafford Act incidents, Federal agencies who have responsibility for on-scene, tactical-level operations may establish incident command and area command structures, or coordinate with state and local agencies to form unified incident command and unified area command structures.

Federal Regional Operational Support

Coordinating structures can be assembled and organized at the regional level to address incidents that cross state borders or have broad geographic or system-wide implications or to manage competing requirements for response assets among multiple incidents.

Federal Regional Facilities

Most Federal departments and agencies have regional or field offices that may participate with state and local governments in planning for incidents and provide response assets when an incident occurs in their jurisdiction. Some Federal departments and agencies share the same standard Federal regional structure as FEMA. In larger-scale incidents, these regional and field offices may provide the initial response assets with additional support being provided from other department and agency offices across the Nation. Some Federal regional and field offices have their own EOCs to support deployments of their assets.

Figure 5. FEMA Regions.

- **FEMA Regional Response Coordination Center (RRCC).** FEMA has 10 regional offices, each headed by a Regional Administrator (see Figure 5). Each of FEMA's regional offices maintains an RRCC. When activated, RRCCs are multi-agency coordination centers generally staffed by ESFs in anticipation of or immediately following an incident. Operating under the direction of the FEMA Regional Administrator, the staff within the RRCCs coordinates Federal regional response efforts and maintains connectivity with FEMA Headquarters and with state EOCs, state and major urban area fusion centers, Federal Executive Boards, and other Federal and state operations and coordination centers that potentially contribute to the development of situational awareness. The UCG assumes responsibility for coordinating Federal response activities at the incident level once Unified Coordination is established, freeing the RRCC to deal with new incidents should they occur.

Federal Headquarters Operational Support

Coordinating structures are assembled and organized at the headquarters level, particularly to address incidents that cross regional borders or have broad geographic or system-wide implications.

Federal Operations Centers

Most Cabinet-level departments and agencies have at least one headquarters-level operations center. A wide range of such centers maintain situational awareness within their functional areas and provide relevant information to the National Operations Center (NOC). These operations centers may also coordinate ESF activities, communicate with other Federal operations centers, and communicate with their local, state, tribal, territorial, and insular area government counterparts. Examples of Federal Operations Centers include:

- **National Operations Center (NOC).** In the event of an act of terrorism, natural disaster, or other emergency, the National Operations Center (NOC),[31] as the principal operations center for the Department of Homeland Security, coordinates and integrates information from NOC components to provide situational awareness and a common operating picture for the entire Federal Government, as well as for local, tribal, and state governments, as appropriate, to ensure that accurate and critical terrorism and disaster-related information reaches government decision makers in a timely manner. Additionally, the NOC serves as the national fusion center, collecting and synthesizing all-source information, including information from state and major urban area fusion centers, for all threats and hazards across the entire integrated national preparedness system.
- **National Response Coordination Center (NRCC).** When activated, the NRCC is a multiagency coordination center located at FEMA Headquarters. Its staff coordinates the overall Federal support for major disasters and emergencies, including catastrophic incidents and emergency management program implementation. FEMA maintains the NRCC as a functional component of the NOC for incident support operations.
- **National Military Command Center (NMCC).** DOD's NMCC is the Nation's focal point for continuous monitoring and coordination of worldwide military operations. It directly supports combatant commanders, the Chairman of the Joint Chiefs of Staff, the Secretary of Defense, and the President in the command of U.S. Armed Forces in peacetime contingencies and war. The NMCC participates in a wide variety of activities, ranging from missile warning and attack assessment to management of peacetime operations such as Defense Support of Civil Authorities during national emergencies.
- **Strategic Information and Operations Center (SIOC).** The SIOC acts as the FBI's worldwide EOC. The SIOC maintains situational awareness of criminal or terrorist threats, critical incidents and crises, both foreign and domestic, regardless of cause or origin, and provides FBI headquarters executives, domestic field offices, and overseas legal attachés with timely notification and the dissemination of strategic information. The SIOC shares information and intelligence with other EOCs at all levels of government. Maintaining a constant state of readiness to support any crisis or major event, the SIOC provides a secure venue to support crisis management, special event monitoring, and significant operations. It provides command, control, communications connectivity, and a common operating picture for managing FBI operational responses and assets anywhere in the world on behalf of FBI Headquarters divisions, field offices, and legal attachés. In the event of a crisis, the SIOC establishes the headquarters command post and develops connectivity to field command posts and Joint Operations Centers.

The specific structures that are activated for any given incident depend on the levels of government involved, as well as the legal authorities under which the response is being conducted.

Integration

Effective emergency response requires the ability for the response coordinating structures to link to and share information with the coordinating structures in the other mission areas. For example, in the wake of a terrorist attack that results in the need for a coordinated Federal response, Response mission area coordinating structures must work closely with those in the Prevention, Protection, Mitigation, and Recovery mission areas. Effective mitigation efforts directly reduce the required scale of response operations. Prevention and protection activities continue after an attack to prevent and protect from follow-on attacks. This requires close coordination of prevention and protection activities with response and recovery efforts. Integration of response mission activities with protection efforts may also occur in the context of a credible threat. Following determination of such a threat, Protection mission area organizations may switch to an enhanced steady-state posture. At that time, Response mission area assets may need to be positioned to respond quickly should protection, mitigation, and prevention efforts fail. Establishing close working relationships, lines of communication, and coordination protocols between protection, prevention, response, and recovery organizations facilitates this process.

Examples of Response mission area coordinating structures cooperating with other mission area assets include:

- Coordinating with Prevention and Protection mission area structures to share threat information including issuing watches, warnings, and other emergency bulletins
- Coordinating with Protection mission area structures in the wake of an incident to ensure that communities and emergency responders have the protection needed to perform their jobs
- Coordinating anticipatory Response mission area activities with the Mitigation and Recovery mission activities.

Although they are generally considered to be prevention or protection focused organizations, the various state and major urban area fusion centers are examples of coordinating structures whose utility spans mission areas. The collection, analysis, and dissemination of information by the fusion centers can inform response activities through information sharing and operational coordination efforts.

Because of the natural relationship between response and recovery efforts and the fact that response and recovery activities often occur simultaneously, the responsibilities of some ESFs overlap with or transition to the responsibilities of Recovery Support Functions (RSFs), the Recovery mission area coordinating structures defined in the National Disaster Recovery Framework. The RSFs frequently build on the ESF resources and short-term recovery efforts applied by the ESFs to meet basic human needs to integrate short-term recovery efforts with intermediate and long-term recovery needs. The relationships and integration between the ESFs and the coordinating structures of other mission areas are detailed in the FIOPs.

RELATIONSHIP TO OTHER MISSION AREAS

All five mission areas integrate with each other through interdependencies, shared assets, and overlapping objectives. These overlapping areas are identified through comprehensive planning with the whole community to ensure that they are addressed during response to an incident.

The Response mission area integrates with the other four mission areas in the following manner:

- **Prevention.** Response organizations coordinate with those responsible for preventing imminent acts of terrorism to understand potential and specific threats and to prepare accordingly by planning for general threats and through crisis action planning for credible threats. Response mission area capabilities must be available in case efforts to prevent terrorist attacks fail or credible threats are identified. Coordinating with prevention officials aids response officials in understanding the extraordinary response capabilities that terrorist attacks may require. When response activities are occurring, whether due to a terrorist attack or another type of incident, prevention activities continue.
- **Protection.** Efforts to protect people and communities, as well as vital facilities, systems, and resources, are inextricably linked to response efforts. Responders that support the Protection and Recovery mission areas include many of the same people and organizations. Protection activities occur before, during, and after incidents. In the aftermath of an incident, a physically secure environment should be established before Response mission area organizations can deliver essential response capabilities.
- **Mitigation.** Reducing risk through hazard mitigation reduces requirements for response capabilities. Mitigation organizations often have special insight into risks and hazards that can be shared with response personnel to improve response planning and execution.
- **Recovery.** As with Protection, the Response and Recovery mission areas include some of the same people and organizations. Communities should build general recovery plans before an incident occurs. After an incident, recovery efforts must begin as soon as possible, often while response capabilities are still being applied.

OPERATIONAL PLANNING

Planning across the full range of homeland security operations is an inherent responsibility of every level of government. This NRF fosters unity of effort for emergency operations planning by providing common doctrine and purpose.

A plan is a continuous, evolving instrument of anticipated actions that maximizes opportunities and guides response operations. Since planning is an ongoing process, a plan is a product based on information and understanding at the moment and is subject to revision.

Operational planning is conducted across the whole community, including the private sector, NGOs, and all levels of government. Comprehensive Preparedness Guide (CPG) 101 provides further information on the various types of plans and guidance on the fundamentals of planning.

From the Federal perspective, integrated planning helps explain how Federal departments and agencies and other national-level whole community partners provide the right resources at the right time to support local, state, tribal, territorial, and insular area government response operations. From their perspectives, integrated planning provides answers to questions about which traditional and non-traditional partners can provide the necessary resources.

The following section outlines how operational planning is applied within the Response mission area and provides guidance for the development of the Response FIOP.

Response Operational Planning

Federal Planning

At the Federal level, the NRF is supported by the Response FIOP. The concepts in the NRF and NIMS guide Federal operational response planning and development of the Response FIOP, which provides further information regarding roles and responsibilities, identifies the critical tasks an entity will take in executing core capabilities, and identifies resourcing and sourcing requirements.

The Response FIOP further defines the concepts, principles, structures, and actions introduced in this Framework with a specific focus on these elements at the Federal level. It addresses interdependencies and integration with the other mission areas throughout the plan's concept of operations. It also describes the management of concurrent actions and coordination points with the areas of prevention, protection, mitigation, and recovery.

The Response FIOP takes an all-hazards approach to preparedness, highlights key areas of interoperability across the five mission areas, and addresses the whole community to optimize resources. The concept of operations in the Response FIOP is based on a no-notice catastrophic incident that spans multiple regions and states and assumes hundreds of thousands of casualties, severe damage to critical infrastructure, and limited ingress and egress due to massive damage to transportation systems. Such an incident would have significant ramifications on the political, economic, social, environmental, logistical, technical, legal, and administrative structures and would overwhelm local, state, tribal, territorial, and insular area government response capabilities.

While the planning factors used for the Response FIOP suggest an incident that will result in a Stafford Act declaration, the plan also addresses the responsibility of certain Federal departments and agencies to lead elements of a response under their own authorities.

The Response FIOP contains:

- A detailed concept of operations
- A description of critical tasks and responsibilities
- Detailed resourcing, personnel, and sourcing requirements
- Specific provisions for the rapid integration of resources and personnel to incidents caused by any of the hazards/threats to which the whole community is particularly vulnerable
- Functional and incident-specific annexes as necessary.

It does not contain detailed descriptions of specific department or agency functions as such information is located in department- or agency-level operational plans.

The NRF is based on the concept of tiered response with an understanding that most incidents start at the local level, and as needs exceed resources and capabilities, additional local, state, and Federal assets are applied. The Response FIOP, therefore, is intended to align with other local, state, tribal, territorial, insular area government, and Federal plans to ensure that all response partners share a common operational focus. Similarly, integration occurs at the Federal level among the departments, agencies, and nongovernmental partners that compose the respective mission area through the frameworks, FIOPs, and departmental and agency operations plans.

In developing the Response FIOP, the following planning needs are taken into account:

- Food and water
- Physically accessible evacuation and sheltering
- Accessible transportation
- Medical surge, medical countermeasures, and treatment capability
- General and medical supplies, durable medical equipment, and disability-related assistance/FNSS
- Emotional, behavioral, and mental health needs
- Reunification and safety of unaccompanied minors
- Guardianship
- Accessible communications
- Animal emergency management needs.

Planning Assumptions

The detailed planning factors for the Response FIOP focus on the impacts associated with a large-scale emergency or disaster that could occur anywhere within the continental United States, its territories, or insular area governments and results in a substantial number of fatalities and injuries, widespread property loss, and disruption of essential services across a large geographic area. Such an occurrence has significant ramifications on the political, economic, social, environmental, logistical, technical, legal, and administrative structures within the impacted area and may overwhelm governmental response capabilities.

The plan addresses the potential, unique requirements and needs of all members of the whole community. While the Response FIOP contains assumptions for each of the response core capabilities, some of the overarching assumptions include the following:

- Multiple catastrophic incidents or attacks will occur with little or no warning
- Incidents are typically managed at the lowest possible geographic, organizational, and jurisdictional level
- Incident management activities will be initiated and conducted using the principles contained in NIMS
- The combined expertise and capabilities of government at all levels, the private sector, and NGOs will be required to respond to a catastrophic incident.

Framework Application

Implementation of the concepts within the NRF and Response FIOP is mandatory for Federal departments and agencies. While the NRF does not direct the actions of other response elements, the guidance contained in the NRF and the Response FIOP is intended to inform local, state, tribal, territorial, and insular area governments, as well as NGOs and the private sector, regarding how the Federal Government responds to incidents. These partners can use this information to inform their planning and ensure that assumptions regarding Federal assistance and response and the manner in which Federal support will be provided are accurate.

SUPPORTING RESOURCES

To assist NRF users, FEMA will maintain an online repository that contains electronic versions of the current NRF documents—base document, ESF Annexes, Support Annexes, and Incident Annexes—as well as other supporting materials. This Resource Center will provide information, training materials, and other tools, such as an overview of the main Stafford Act provisions, a guide to authorities and references, and an abbreviation list to assist response partners in understanding and executing their roles under the NRF.

Resource Center materials will be regularly evaluated, updated, and augmented as necessary. Additional content may be added or modified at the request of Response mission area partners and other users.

CONCLUSION

The environment in which the Nation operates grows ever more complex and unpredictable. In implementing the NRF to build national preparedness, partners are encouraged to develop a shared understanding of broad-level strategic implications as they make critical decisions in building future capacity and capability. The whole community should be engaged in examining and implementing the strategy and doctrine contained in this Framework, considering both current and future requirements in the process. This means that this Framework is a living document, and it will be regularly reviewed to evaluate consistency with existing and new policies, evolving conditions, and the experience gained from its use. The first review will be completed no later than 18 months after the release of the Framework. Subsequent reviews will be conducted in order to evaluate the effectiveness of the Framework on a quadrennial basis.

DHS will coordinate and oversee the review and maintenance process for the NRF. The revision process includes developing or updating any documents necessary to carry out capabilities. Significant updates to the Framework will be vetted through a Federal senior-level interagency review process. This Framework will be reviewed in order to accomplish the following:

- Assess and update information on the core capabilities in support of Response goals and objectives
- Ensure that it adequately reflects the organization of responsible entities
- Ensure that it is consistent with the other four mission areas
- Update processes based on changes in the national threat/hazard environment
- Incorporate lessons learned and effective practices from day-to-day operations, exercises, and actual incidents and alerts
- Reflect progress in the Nation's Response mission activities, the need to execute new laws, executive orders, and Presidential directives, as well as strategic changes to national priorities and guidance, critical tasks, or national capabilities.

The implementation and review of this Framework will consider effective practices and lessons learned from exercises and operations, as well as pertinent new processes and technologies. Effective practices include continuity planning, which ensures that the capabilities contained in this Framework can continue to be executed regardless of the threat or hazard. Pertinent new processes and technologies should enable the Nation to adapt efficiently to the evolving risk environment and use data relating to location, context, and interdependencies that allow for effective integration across all missions using a standards-based approach. Updates to the NRF Annexes may occur independently from reviews of the base document.

America's security and resilience work is never finished. While the Nation is safer, stronger, and better prepared than it was a decade ago, the commitment to safeguard the Nation against its greatest risks, now and for decades to come, remains resolute. By bringing the whole community together now to address future needs, the Nation will continue to improve its preparedness to face whatever challenges unfold.

End Notes

[1] As with all activities in support of the National Preparedness Goal, activities taken under the response mission must be consistent with all pertinent statutes and policies, particularly those involving privacy and civil and human rights, such as the Americans with Disabilities Act of 1990, Rehabilitation Act of 1973, and Civil Rights Act of 1964.

[2] http://www.fema.gov/emergency/nims

[3] A catastrophic incident is defined as any natural or manmade incident, including terrorism, that results in extraordinary levels of mass casualties, damage, or disruption severely affecting the population, infrastructure, environment, economy, national morale, or government functions.

[4] Per the Stafford Act, insular areas include Guam, the Commonwealth of the Northern Mariana Islands, American Samoa, and the U.S. Virgin Islands. Other statutes or departments and agencies may define the term insular area differently.

[5] Whole community includes: individuals, families, households, communities, the private and nonprofit sectors, faith-based organizations, and local, state, tribal, territorial, and Federal governments. Whole community is defined in the National Preparedness Goal as "a focus on enabling the participation in national preparedness activities of a wider range of players from the private and nonprofit sectors, including nongovernmental organizations and the general public, in conjunction with the participation of Federal, state, and local governmental partners in order to foster better coordination and working relationships." The National Preparedness Goal may be found online at http://www.fema.gov/ppd8.

[6] Access and functional needs includes ensuring the equal access and meaningful participation of all individuals, without discrimination.

[7] For further information, see the Core Capabilities section.

[8] A domestic incident may have international and diplomatic impacts and implications that call for coordination and consultations with foreign governments and international organizations. The NRF also applies to the domestic response to incidents of foreign origin that impact the United States. See the International Coordination Support Annex for more information.

[9] As with all activities in support of the National Preparedness Goal, activities taken under the response mission must be consistent with all pertinent statutes and policies, particularly those involving privacy and civil and human rights, such as the Americans with Disabilities Act of 1990, Rehabilitation Act of 1973, and Civil Rights Act of 1964.

[10] Information, warnings, and communications associated with emergency management must ensure effective communication, such as through the use of appropriate auxiliary aids and services (e.g., interpreters, captioning, alternate format documents), for individuals with disabilities and provide meaningful access to limited English proficient individuals.

[11] Certain incidents such as a pandemic or cyber event may not be limited to a specific geographic area and may be managed at the local, state, tribal, territorial, insular area, or Federal level depending on the nature of the incident.

[12] The Incident Command System's "unified command" concept is distinct from the military use of this term. Concepts of "command" and "unity of command" have distinct legal and cultural meanings for military forces and military operations. Military forces always remain under the control of the military chain of command and are subject to redirection or recall at any time. Military forces do not operate under the command of the incident commander or under the unified command structure, but they do coordinate with response partners and work toward a unity of effort while maintaining their internal chain of command.

[13] Additional information is available at http://www.nvoad.org.

[14] Additional information on the protection of critical infrastructure can be found in the National Protection Framework, the National Infrastructure Protection Plan, and the NRF Critical Infrastructure Support Annex.

[15] States are sovereign entities, and the governor has responsibility for public safety and welfare. Although U.S. territories, possessions, freely associated states, and tribal governments also have sovereign rights, there are unique factors involved in working with these entities. Stafford Act assistance is available to states and to the District of Columbia, Puerto Rico, the Virgin Islands, Guam, American Samoa, and the Commonwealth of the Northern Mariana Islands, which are included in the definition of "state" in the Stafford Act. Federal disaster preparedness, response, and recovery assistance is available to the Federated States of Micronesia and the Republic of the Marshall Islands pursuant to Compacts of Free Association. The extent to which Federal response or assistance is provided to insular areas, territories, and tribes under other Federal laws is defined in those laws and supporting regulations.

[16] "Governor" is used throughout this document to refer to the chief executive of states, territories, and insular areas.

[17] A reference paper on EMAC is available at http://www.emacweb.org.

[18] The President may call National Guard forces into Federal service for domestic duties under Title 10 (e.g., in cases of invasion by a foreign nation, rebellion against the authority of the United States, or where the President is unable to execute the laws of the United States with regular forces under 10 U.S. Code §12406). When National Guardsmen are employed under Title 10 of the U.S. Code, these forces are no longer under the command of the governor. Instead, the Department of Defense assumes full command and control of National Guard forces called into Federal service.

[19] Except for those activities that may interfere with the authority of the Attorney General or the FBI Director, as described in PPD-8.

[20] A declaration of a public health emergency may make available any funds appropriated to the Public Health Emergency Fund.

[21] The Commandant of the USCG coordinates the designation of a Spill of National Significance with the Secretary of Homeland Security, as appropriate.

[22] See the ESF #10 – Oil and Hazardous Materials Response Annex for more information on these authorities.

[23] See the Post-Katrina Emergency Management Reform Act, enacted as part of the FY 2007 DHS Appropriations Act, P.L. 109-295.

[24] In response to a request for assistance from a civilian authority, under imminently serious conditions, and if time does not permit approval from higher authority, DOD officials may provide an immediate response by temporarily employing the resources under their control, subject to any supplemental direction provided by higher headquarters, to save lives, prevent human suffering, or mitigate great property damage within the United States. Immediate response authority does not permit actions that would subject civilians to the use of military power that is regulatory, prescriptive, proscriptive, or compulsory. (DOD Directive 3025.18)

[25] Given the scope and magnitude of a catastrophic incident, waivers, exceptions, and exemptions to policy, regulations, and laws may be available in order to save and sustain life, and to protect property and the environment. However, any such waivers, exceptions, and exemptions must be consistent with laws that preserve human and civil rights and protect individuals with disabilities and others with access and functional needs.

[26] The critical infrastructure sectors are described in the National Infrastructure Protection Plan.

[27] Individual and household preparedness information can be located at http://www.ready.gov/make-a-plan.

[28] Additional information sharing and collaborative opportunities can be located at FEMA Private Sector Focus http://www.fema.gov/privatesector/index.shtm.

[29] The President has delegated most of his authority under the Stafford Act to the Secretary of Homeland Security, who has, In turn, delegated those authorities to the FEMA Administrator.

[30] Refer to footnote 15 for more information on U.S. possessions and freely associated states.

[31] The NOC is more than a brick and mortar command center. It is composed of the NOC Watch, Intelligence Watch and Warning, FEMA National Watch Center and National Response Coordination Center, and the National Infrastructure Coordinating Center. The DHS National Cybersecurity and Communications Integration Center also maintains and updates the NOC regarding the national common operational picture for cyberspace.

In: National Response and Disaster Recovery Frameworks ISBN: 978-1-63321-892-5
Editor: Daniella Beckett

Chapter 2

NATIONAL DISASTER RECOVERY FRAMEWORK*

Federal Emergency Management Agency

1. EXECUTIVE SUMMARY

Experience with recent disaster recovery efforts highlights the need for additional guidance, structure and support to improve how we as a Nation address recovery challenges. This experience prompts us to better understand the obstacles to disaster recovery and the challenges faced by communities that seek disaster assistance. The *National Disaster Recovery Framework (NDRF)* is a guide to promote effective recovery, particularly for those incidents that are large- scale or catastrophic.

The NDRF provides guidance that enables effective recovery support to disaster-impacted States, Tribes and local jurisdictions. It provides a flexible structure that enables disaster recovery managers to operate in a unified and collaborative manner. It also focuses on how best to restore, redevelop and revitalize the health, social, economic, natural and environmental fabric of the community and build a more resilient Nation.

The NDRF defines:

- Core recovery *principles*
- *Roles and responsibilities* of recovery coordinators and other stakeholders
- A *coordinating structure* that facilitates communication and collaboration among all stakeholders
- Guidance for pre- and post-disaster recovery *planning*
- The overall *process* by which communities can capitalize on opportunities to rebuild stronger, smarter and safer

* This is an edited, reformatted and augmented version of a document issued September 2011.

These elements improve recovery support and expedite recovery of disaster-impacted individuals, families, businesses and communities. While the NDRF speaks to all who are impacted or otherwise involved in disaster recovery, it concentrates on support to individuals and communities.

The NDRF introduces four new concepts and terms:

- Federal Disaster Recovery Coordinator (FDRC)
- State or Tribal Disaster Recovery Coordinators (SDRC or TDRC)
- Local Disaster Recovery Managers (LDRM)
- Recovery Support Functions (RSFs)

The FDRC, SDRC, TDRC and LDRM provide focal points for incorporating recovery considerations into the decisionmaking process and monitoring the need for adjustments in assistance where necessary and feasible throughout the recovery process. The RSFs are six groupings of core recovery capabilities that provide a structure to facilitate problem solving, improve access to resources, and foster coordination among State and Federal agencies, nongovernmental partners and stakeholders. Each RSF has coordinating and primary Federal agencies and supporting organizations that operate together with local, State and Tribal government officials, nongovernmental organizations (NGOs) and private sector partners. The concepts of the FDRCs, SDRCs, TDRCs and RSFs are scalable to the nature and size of the disaster.

The NDRF aligns with the *National Response Framework (NRF)*. The NRF primarily addresses actions during disaster response. Like the NRF, the NDRF seeks to establish an operational structure and to develop a common planning framework. The NDRF replaces the NRF Emergency Support Function #14 (ESF #14) - Long-Term Community Recovery. Key ESF #14 concepts are expanded in the NDRF and include recovery-specific leadership, organizational structure, planning guidance and other components needed to coordinate continuing recovery support to individuals, businesses and communities.

Fundamentally, the NDRF is a construct to optimally engage existing Federal resources and authorities, and to incorporate the full capabilities of all sectors in support of community recovery. The effective implementation of the NDRF, whether or not in the context of a *Robert T. Stafford Disaster Relief and Emergency Assistance Act (Stafford Act)* declaration, requires strong coordination across all levels of government, NGOs and the private sector. It also requires an effective, accessible public information effort so that all stakeholders understand the scope and the realities of recovery. The NDRF provides guidance to assure that recovery activities respect the civil rights and civil liberties of all populations and do not result in discrimination on account of race, color, national origin (including limited English proficiency), religion, sex, age or disability. Understanding legal obligations and sharing best practices when planning and implementing recovery strategies to avoid excluding groups on these bases is critical.

The NDRF is a guide to promote effective recovery. It is a concept of operations and not intended to impose new, additional or unfunded net resource requirements on Federal agencies. As responsibilities, capabilities, policies and resources expand or change, the NDRF will be revised as needed to ensure that it continues to provide a common and adaptable approach to disaster recovery.

2. INTRODUCTION

The *National Disaster Recovery Framework (NDRF)* describes the concepts and principles that promote effective Federal recovery assistance. It identifies scalable, flexible and adaptable coordinating structures to align key roles and responsibilities. It links local, State, Tribal and Federal governments, the private sector and nongovernmental and community organizations that play vital roles in recovery. The NDRF captures resources, capabilities and best practices for recovering from a disaster. It recognizes that significant challenges confront all recovery efforts, from a relatively localized incident to a large-scale disaster that demands substantial resources. Importantly, the NDRF is intended to address disasters of all kinds and sources, whether it is a major Presidentially-declared disaster or a non-Presidentially declared incident. The NDRF is a companion document to the *National Response Framework (NRF)* and is supported by the ongoing development of detailed operational, management, field guidance and training tools.

In September 2009, President Barack Obama charged the U.S. Department of Homeland Security (DHS) and the U.S. Department of Housing and Urban Development (HUD) to establish a Long- Term Disaster Recovery Working Group (the Working Group). Composed of more than 20 Federal departments, agencies and offices, the Working Group was asked to develop operational guidance for recovery organizations, which resulted in the creation of the NDRF, and to make recommendations for improving the Nation's approach to disaster recovery.

During the fall of 2009, DHS/Federal Emergency Management Agency (FEMA) and HUD sponsored outreach sessions in each of FEMA's ten regions and stakeholder forums in five cities across the country. The objective was to offer stakeholders from a wide array of organizations and backgrounds the opportunity to provide up-front comments to the Working Group on ways to strengthen disaster recovery. DHS/FEMA and HUD also organized discussion roundtables with professional associations and academic experts. The Working Group created a Web portal, which enabled a large and diverse group of stakeholders to provide comments into the development of the NDRF. Over six hundred stakeholders representing local, State, Tribal and Federal governments, as well as public and private sector organizations from across the Nation contributed more than six thousand comments.

The NDRF reflects as core principles nine significant themes and recommendations that emerged from these stakeholder outreach efforts. These principles are:

- Individual and Family Empowerment
- Leadership and Local Primacy
- Pre-Disaster Recovery Planning
- Partnerships and Inclusiveness
- Public Information
- Unity of Effort
- Timeliness and Flexibility
- Resilience and Sustainability
- Psychological and Emotional Recovery

Built as a document to forge a common understanding of roles, responsibilities and resources available for effective recovery, the NDRF is designed for anyone who is involved in disaster recovery. Key concepts in the document are the need for:

- **Structure** — Provided by Recovery Support Functions (RSFs).
- **Leadership** — Provided locally and strengthened through support by the State or Tribal Disaster Recovery Coordinator (SDRCs or TDRCs); Local Disaster Recovery Managers (LDRMs); RSFs; private sector and nongovernmental organization (NGO) leaders; and when needed, the Federal Disaster Recovery Coordinator (FDRC).
- **Planning** — Developed during both pre- and post-disaster phases.

These concepts are explained and developed in the NDRF. When combined with the full involvement of all stakeholders, along with realistic and well-communicated expectations of desired outcomes, the concepts constitute the building blocks for a successful recovery.

The NDRF and supporting guidance and tools that follow its publication form the framework of a national disaster recovery strategy.

3. Purpose of the Framework

The *National Disaster Recovery Framework (NDRF)* defines how Federal agencies will more effectively organize and operate to utilize existing resources to promote effective recovery and support States, Tribes and other jurisdictions affected by a disaster. It is also written for a larger audience of non-Federal Government executives, private sector and nongovernmental organization (NGO) leaders, emergency managers, community development professionals and disaster recovery practitioners[1].

Recovery begins with pre-disaster preparedness and includes a wide range of planning activities. The NDRF clarifies the roles and responsibilities for stakeholders in recovery, both pre- and post-disaster. It recognizes that recovery is a continuum and that there is opportunity within recovery. It also recognizes that when a disaster occurs, it impacts some segments of the population more than others.

The ability of a community to accelerate the recovery process begins with its efforts in pre-disaster preparedness, mitigation and recovery capacity building. These efforts result in a resilient community with an improved ability to withstand, respond to and recover from disasters. Timely decisions in response to disaster impacts can significantly reduce recovery time and cost.

The NDRF describes key principles and steps for community recovery planning and implementation. It promotes a process in which the impacted community fully engages and considers the needs of all its members. A key element of the process is that the impacted community assumes the leadership in developing recovery priorities and activities that are realistic, well-planned and clearly communicated.

The NDRF advances the concept that recovery encompasses more than the restoration of a community's physical structures to its pre- disaster conditions. Of equal importance is providing a continuum of care to meet the needs of the affected community members who have experienced the hardships of financial, emotional or physical impacts as well as positioning the community to meet the needs of the future. The NDRF also highlights the importance of disaster recovery activities that promote sustainability practices. These practices may reduce community vulnerability to recurrent disasters. Meeting these various needs — through strengthening the health and human services, social fabric, educational system, environmental sustainability, cultural resources and economic vitality — serves to enhance the overall resiliency of the entire community as the recovery progresses.

Resources

The *National Disaster Recovery Framework (NDRF)* is a guide to promote effective recovery — it is a concept of operations and not intended to impose new, additional or unfunded net resource requirements on Federal agencies. Instead, the NDRF aims to leverage and concentrate the effects of existing Federal resources, programs, projects and activities through an organization of Recovery Support Functions (RSFs) to promote effective recovery for affected communities before and after disaster strikes. The National Disaster Recovery Planning (NDRP) Division at Federal Emergency Management Agency (FEMA) Headquarters facilitates and coordinates RSF activities and recovery planning at the national level. Each RSF coordinating agency will commit to designating a senior level principal to serve as the RSF national coordinator, provide significant engagement and management for the RSF, and ensure ongoing communication and coordination between the primary agencies and support organizations for the RSFs. The RSF national coordinator also ensures coordination and communication between the Federal agencies and corresponding local, State and Tribal authorities and nongovernmental and private-sector organizations throughout the preparedness, response and recovery phases of a disaster.

The NDRF is not intended to increase overall Federal agency activity in support of recovery planning during steady-state. Accordingly, Federal agencies with NDRF roles and responsibilities shall fund the costs arising from those responsibilities out of their base budgets and staffing levels, and, except as noted above, shall only support steady-state NDRF activities subject to available resources. The operational costs of Federal recovery programs will continue to be borne by agencies from appropriations made for such purposes, except for those expenses authorized for reimbursement under the *Robert T. Stafford Disaster Relief and Emergency Assistance Act (Stafford Act)* or as otherwise provided by law.

Applicability

The National Disaster Recovery Framework (NDRF) applies to all Presidentially-declared major disasters though not all elements will be activated for every declared incident. Many of its concepts and principles are equally valid for non-declared incidents that have recovery consequences. The core concepts as well as the Recovery Support Function (RSF) organizing structures outlined in the NDRF may be applied to any incident regardless of whether or not it results in a Presidential disaster declaration.

Similar to how the *National Response Framework (NRF)* is the overarching interagency response coordination structure for both *Robert T. Stafford Disaster Relief and Emergency Assistance Act (Stafford Act)* and non-Stafford Act incidents, the NDRF will provide the overarching interagency coordination structure for the recovery phase for Stafford Act incidents, and elements of the framework may also be used for significant non-Stafford Act incidents. For example, the Federal response to an oil "Spill of National Significance," as defined under the *National Oil and Hazardous Substances Pollution Contingency Plan*, more commonly known as the *National Contingency Plan (NCP)*, may be managed under the NCP without a Stafford Act declaration. Elements of the NDRF also may be activated as needed to provide coordinated Federal recovery assistance. The response to the 2010 Deepwater Horizon Oil Spill was an example of an oil Spill of National Significance that was managed under the NCP, and further supplemented by additional Federal recovery assistance.

Relationship to the National Response Framework

The focus of the *National Response Framework (NRF)* is the response actions as well as the short-term recovery activities that immediately follow or overlap those actions. The *National Disaster Recovery Framework (NDRF)* does not speak to these short-term activities such as life saving, life sustaining, property protection and other measures intended to neutralize the immediate threat to life, environment and property, as well as to stabilize the community. However, these activities influence recovery activities, necessitating the need for a structure to consider and advise on recovery implications during the early phases of incident management. The NDRF provides the tools to encourage early integration of recovery considerations into the response phase operations.

As response, short-term and intermediate recovery activities begin to wind down, recovery needs gradually take on a more critical role. The core principles and organizational constructs introduced in the NDRF coexist with the NRF and build upon its organizational structure and resources to more effectively address recovery needs. The NRF fully transitions to the NDRF when the disaster-specific mission objectives of the Emergency Support Functions (ESFs) are met and all ESFs demobilize.

Together, the NDRF and the NRF provide the doctrine and guidance to implement the response and recovery aspects of the *National Homeland Security Strategy (2007)*. In addition, the *National Infrastructure Protection Plan (NIPP)* and the *Critical Infrastructure and Key Resources (CIKR) Annex* to the NRF provide a bridge between steady-state CIKR protection and response and recovery programs designed to support the maintenance and restoration of the Nation's CIKR. These documents incorporate and adopt the central tenets of the *National Incident Management System (NIMS)* and support the primacy of local, State and Tribal governments in preparing for and managing the response and recovery from natural and human-caused disasters.

National Preparedness System

The *National Disaster Recovery Framework (NDRF)* will be revised as the National Preparedness System is further developed and the Prevention, Protection, Mitigation and Response Frameworks are completed or updated to ensure that actions taken in the NDRF are coordinated with relevant actions described in the other frameworks across the preparedness spectrum. In addition, core recovery capabilities will be further defined as interagency operational plans and planning guidance documents are developed to support the NDRF as part of the National Preparedness System.

Recovery Continuum

The recovery process is best described as a sequence of interdependent and often concurrent activities that progressively advance a community toward a successful recovery. However, decisions made and priorities set early in the recovery process by a community will have a cascading effect on the nature and speed of the recovery progress. Figure 1 indicates how response and recovery functions are related in example sectors.

4. Core Principles

Core Principles That Guide Recovery

The *National Disaster Recovery Framework (NDRF)* is guided by nine core principles that, when put into practice, maximize the opportunity for achieving recovery success.

Recovery Core Principles

Individual and Family Empowerment.
Leadership and Local Primacy.
Pre-Disaster Recovery Planning. Partnerships and Inclusiveness.
Public Information.
Unity of Effort.
Timeliness and Flexibility.
Resilience and Sustainability.
Psychological and Emotional Recovery.

Individual and Family Empowerment

All community members must have equal opportunity to participate in community recovery efforts in a meaningful way. Care must be taken to assure that actions, both intentional and unintentional, do not exclude groups of people based on race, color, national origin (including limited English proficiency), religion, sex or disability. Care must be taken to identify and eradicate social and institutional barriers that hinder or preclude individuals with disabilities and others in the community historically subjected to unequal treatment from full and equal enjoyment of the programs, goods, services, activities, facilities, privileges, advantages and accommodations provided. A successful recovery is about the ability of individuals and families to rebound from their losses in a manner that sustains their physical, emotional, social and economic well-being. The restoration of infrastructure systems and services is critical during recovery. It is vital that all individuals who make up the community are provided with the tools to access and use a continuum of care that addresses both the physical losses sustained and the psychological and emotional trauma experienced.

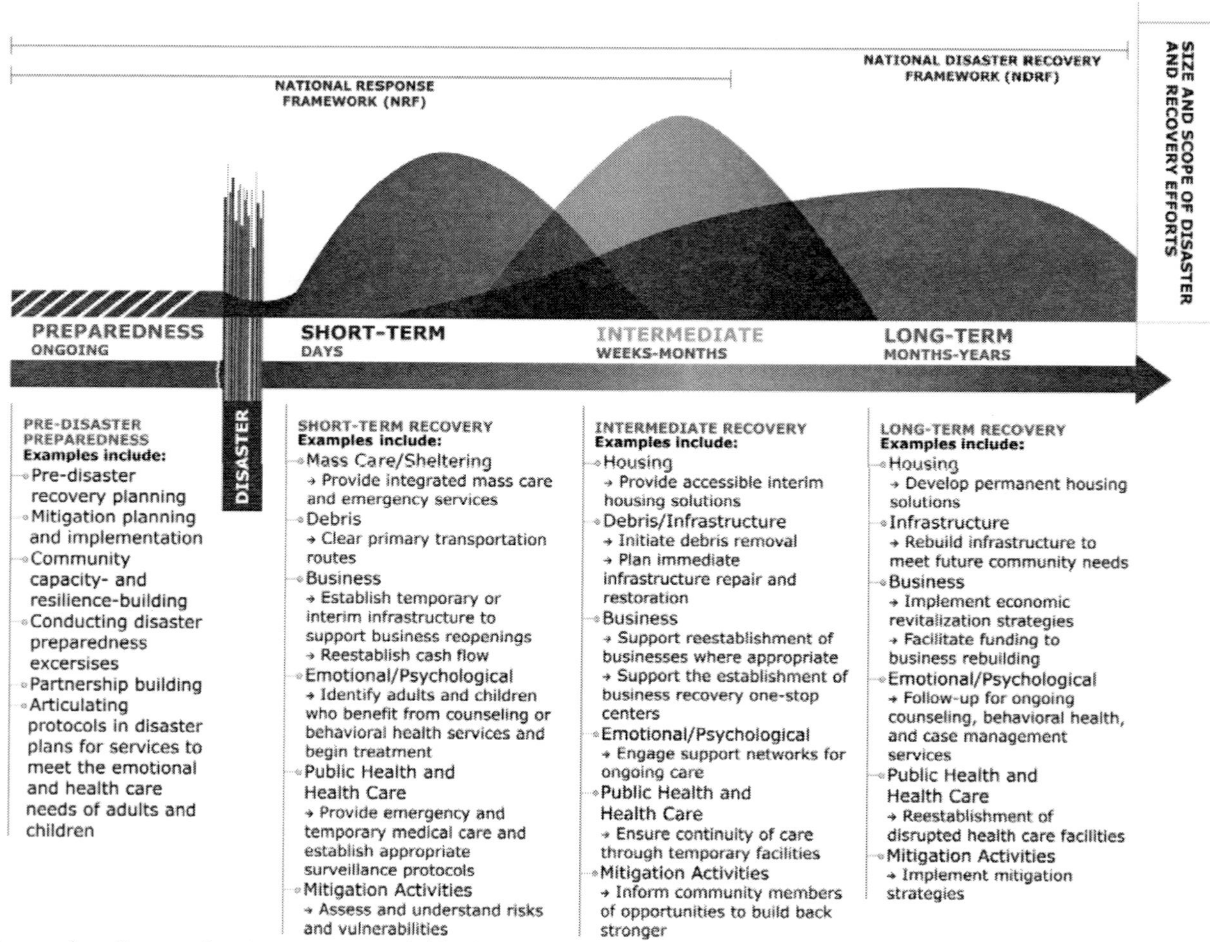

This recovery continuum describes overlapping recovery activities by phase.

Figure 1. Recovery Continuum – Description of Activities by Phase.

Leadership and Local Primacy

Successful recovery requires informed and coordinated leadership throughout all levels of government, sectors of society and phases of the recovery process. It recognizes that local, State and Tribal governments have primary responsibility for the recovery of their communities and play the lead role in planning for and managing all aspects of community recovery. This is a basic, underlying principle that should not be overlooked by State, Federal and other disaster recovery managers. States act in support of their communities, evaluate their capabilities and provide a means of support for overwhelmed local governments. The Federal Government is a partner and facilitator in recovery, prepared to enlarge its role when the disaster impacts relate to areas where Federal jurisdiction is primary or affects national security. The Federal Government, while acknowledging the primary role of local, State and Tribal governments, is prepared to vigorously support local, State and Tribal governments in a large-scale disaster or catastrophic incident.

Pre-Disaster Recovery Planning

The speed and success of recovery can be greatly enhanced by establishment of the process and protocols prior to a disaster for coordinated post-disaster recovery planning and implementation. All stakeholders should be involved to ensure a coordinated and comprehensive planning process, and develop relationships that increase post-disaster collaboration and unified decisionmaking. Another important objective of pre-disaster recovery planning is to take actions that will significantly reduce disaster impacts through disaster-resilient building practices. The NDRF strongly encourages innovation among the States, Tribes, localities, and the private sector in working together to identify State, Tribal and locally- generated tools and resources, pre-disaster, that will serve to support and sustain disaster mitigation and recovery efforts.

Partnerships and Inclusiveness

Partnerships and collaboration across groups, sectors and governments promote a successful recovery process. Partnerships and inclusiveness are vital for ensuring that all voices are heard from all parties involved in disaster recovery and that all available resources are brought to the table. This is especially critical at the community level where nongovernmental partners in the private and nonprofit sectors play a critical role in meeting local needs. Inclusiveness in the recovery process includes individuals with disabilities and others with access and functional needs, advocates of children, seniors and members of underserved populations. Sensitivity and respect for social and cultural diversity must be maintained at all times. Compliance with equal opportunity and civil rights laws must also be upheld.

Public Information

Clear, consistent, culturally appropriate and frequent communication initiatives promote successful public information outcomes. These incorporate a process that is inclusive and ensures accessibility to all, including those with disabilities, persons who are deaf or blind and those with limited English proficiency. Public information messaging helps manage expectations throughout the recovery process and supports the development of local, State and Tribal government communications plans. This ensures stakeholders have a clear understanding of available assistance and their roles and responsibilities; makes clear the actual pace, requirements and time needed to achieve recovery; and includes information and referral help lines and websites for recovery resources.

Unity of Effort

A successful recovery process requires unity of effort, which respects the authority and expertise of each participating organization while coordinating support of common recovery objectives. Common objectives are built upon consensus and a transparent and inclusive planning process with clear metrics to measure progress.

Timeliness and Flexibility

A successful recovery process upholds the value of timeliness and flexibility in coordinating and efficiently conducting recovery activities and delivering assistance. It also minimizes delays and loss of opportunities. The process strategically sequences recovery decisions and promotes coordination; addresses potential conflicts; builds confidence and ownership of the recovery process among all stakeholders; and ensures recovery plans, programs, policies and practices are adaptable to meet unforeseen, unmet and evolving recovery needs.

Resilience and Sustainability

A successful recovery process promotes practices that minimize the community's risk to all hazards and strengthens its ability to withstand and recover from future disasters, which constitutes a community's resiliency. A successful recovery process engages in a rigorous assessment and understanding of risks and vulnerabilities that might endanger the community or pose additional recovery challenges. The process promotes implementation of the *National Infrastructure Protection Plan (NIPP)* risk management framework to enhance the resilience and protection of critical infrastructure against the effects of future disasters. Resilience incorporates hazard mitigation and land use planning strategies; critical infrastructure, environmental and cultural resource protection; and sustainability practices to reconstruct the built environment, and revitalize the economic, social and natural environments.

Psychological and Emotional Recovery

A successful recovery process addresses the full range of psychological and emotional needs of the community as it recovers from the disaster through the provision of support, counseling, screening and treatment when needed. These needs range from helping individuals to handle the shock and stress associated with the disaster's impact and recovery challenges, to addressing the potential for and consequences of individuals harming themselves or others through substance, physical and emotional abuses. Successful recovery acknowledges the linkages between the recovery of individuals, families and communities.

5. ACHIEVING DISASTER RECOVERY

Each community defines successful recovery outcomes differently based on its circumstances, challenges, recovery vision and priorities. One community may characterize success as the return of its economy to pre-disaster conditions while another may see success as the opening of new economic opportunities. Although no single definition fits all situations, successful recoveries do share conditions in which:

- The community successfully overcomes the physical, emotional and environmental impacts of the disaster.
- It reestablishes an economic and social base that instills confidence in the community members and businesses regarding community viability.
- It rebuilds by integrating the functional needs of all residents and reducing its vulnerability to all hazards facing it.
- The entire community demonstrates a capability to be prepared, responsive, and resilient in dealing with the consequences of disasters.

Recovery is more than the community's return to pre-disaster circumstances, especially when the community determines that these circumstances are no longer sustainable, competitive or functional as shown by the community's post-disaster condition. A successful recovery in this case may include a decision to relocate all or some portion of the community assets and restoration of the affected area to a more natural environment. In these circumstances, the community recovery decisionmaking is informed by evaluating all alternatives and options and avoiding simple rebuilding or reconstructing of an area that continues to be vulnerable.

Factors of a Successful Recovery

Experience shows that the presence of certain factors in a community can help ensure a successful recovery.

SUCCESS FACTORS

Effective Decisionmaking and Coordination.
Integration of Community Recovery Planning Processes.
Well-managed Recovery.
Proactive Community Engagement, Public Participation and Public Awareness.
Well-administered Financial Acquisition.
Organizational Flexibility.
Resilient Rebuilding.

Effective Decisionmaking and Coordination

- Recovery leadership defines roles and responsibilities for all stakeholders.
- Businesses, nonprofits and local community leadership examine recovery alternatives, address conflicts and make informed and timely decisions that best achieve recovery of the impacted community.

- Organizations providing leadership or assistance for recovery establish metrics for tracking progress, ensuring accountability and reinforcing realistic expectations among stakeholders.
- Governments, voluntary, faith-based and community organizations provide assistance to track progress, ensure accountability and make adjustments to ongoing assistance.

Integration of Community Recovery Planning Processes

- Communities engage in pre-disaster recovery planning and other recovery preparedness, mitigation and community resilience-building work.
- Individual, business and community preparation and resilience-building provide a foundation for recovery plans that improve the speed and quality of post-disaster recovery decisions.
- The public-private partnership under the *National Infrastructure Protection Plan (NIPP)* facilitates broad coordination and information sharing among all levels of government and private sector owners and operators of critical infrastructure.
- The community develops processes and criteria for identifying and prioritizing key recovery actions and projects.
- The community's recovery leadership creates an organizational framework involving key sectors and stakeholders to manage and expedite recovery planning and coordination.
- Recovery authorities revise existing local and State level emergency response contingencies to include recovery planning best practices and other preparedness, mitigation and community resilience-building work.

Well-Managed Recovery

- Well-established, pre-disaster partnerships at the local, State, Tribal and Federal levels, including those with the private sector and nongovernmental organizations (NGOs), help to drive a successful recovery.
- Recovery stakeholders leverage and coordinate disaster and traditional public and NGO assistance programs to accelerate the recovery process and avoid duplication of efforts.
- Communities seek out, interface and coordinate successfully with outside sources of help, such as surrounding governments, foundations, universities, nonprofit organizations and private sector entities — a key element in rapid recovery.
- Readily available surge staffing and management structures support the increased workload during recovery, such as code enforcement, planning, communications, grant-writing and management.
- Recovery leadership establishes guidance, including the shift of roles and responsibilities, for the transition from response operations, to recovery and finally, a return to a new normal state of community functioning.
- Well-managed recoveries ensure compliance with architectural standards and programmatic accessibility during recovery.

Proactive Community Engagement, Public Participation and Public Awareness

- Stakeholders collaborate to maximize the use of available resources to rebuild housing, infrastructure, schools, businesses and the social-historical- cultural fabric of the impacted community in a resilient manner; and to provide health care, access and functional support services.
- All community perspectives are represented in all phases of disaster and recovery planning; transparency and accountability in the process are clearly evident.
- Communities create post-disaster recovery plans that can be implemented quickly. Local opinions are incorporated so that community needs are met in a more holistic manner, maximizing the provision and utilization of recovery resources and built upon, or incorporated into, the community master plan.
- Public information is accessible to keep everyone informed throughout the recovery process. This includes providing appropriate aids and services, such as captioning, large print, Braille, interpretation and translated materials, to ensure effective communication with individuals with disabilities and to facilitate access to information for individuals with limited English proficiency.
- Continuous and accessible public information campaigns to community members on various recovery programs and the commitment to short, intermediate and long-term recovery, as well as the overall recovery progress, increase public confidence.

Well-Administered Financial Acquisition

- Community stakeholders need to possess an understanding and have access to broad and diverse funding sources in order to finance recovery efforts.
- The community's knowledge and professional administration of external programs greatly aid the recovery progress.
- Funders and resource providers collaborate to provide program flexibility and implement finance planning. Recovery management and program administration collaborate in a post- disaster environment.
- Recovery management programs support the development and maintenance of adequate financial monitoring and accounting systems for new and large levels of investment. Management programs include systems that detect and deter fraud, waste and abuse.
- Federal recovery expenditures maximize the use of local businesses to promote local economic development.

Organizational Flexibility

- Organizational structures for coordinating recovery assistance are scalable and flexible.
- Recovery structures at all government levels evolve, adapt and develop new skills and capacities to address the changing landscape of post-disaster environments.
- Functional and effective intergovernmental relations influence the efficiency of the recovery process.
- Organizational flexibility facilitates the application of laws, regulations and policies in the context of disaster and enhances the government's adaptability to govern in unforeseen incidents.

- Flexible staffing and management structures enhance the adaptability of the governmental structure.
- Increased pre-disaster partnerships help reduce or avoid the challenges of establishing new partnerships in a post- disaster environment.
- Organizational flexibility is compatible with the integrity and accountability of taxpayer-funded programs.

Resilient Rebuilding

- The community rebuilds a sustainable future inclusive of ecological, economic and local capacity considerations.
- The recovery is an opportunity for communities to rebuild in a manner which reduces or eliminates risk from future disasters and avoids unintended negative environmental consequences.
- Communities incorporate stronger building codes and land use ordinances. Vulnerable structures are retrofitted, elevated or removed from harm.
- Community members, businesses and local governments incorporate risk- reduction strategies into governance and local decisionmaking.

Measuring Recovery Progress

Measuring and communicating the progress of recovery increases public confidence in the recovery process by promoting transparency, accountability and efficiency. It enables local leadership to identify ongoing recovery needs and engages partners in providing assistance and problem resolution. Recovery progress serves as a tracking mechanism for improving and adjusting recovery strategies and activities and ensuring continuing improvement. Communities determine how to qualify and quantify their progress. They measure progress toward recovery holistically, recognizing that recovery outcomes and impacts are measured beyond a single criterion such as dollars spent or assistance delivered on a program-by- program basis. The following are successful strategies for measuring progress:

- Recognize that recovery progress has variables not attributable to any one program or government agency. Overall recovery success depends upon the interaction of a wide range of public, nonprofit and private programs and initiatives, good planning, local capacity, leadership, effective decisionmaking and the building of public confidence.
- Establish systems that track pre-disaster baseline conditions, overall recovery of individuals as well as the reconstruction and redevelopment of infrastructure, economy, health, social and community services and government functions.
- Ensure disaster preparedness and recovery planning is integrated with community-wide comprehensive and hazard mitigation planning to capitalize on opportunities that minimize the risk to all hazards and strengthen the ability to withstand and recover from future disasters.
- Select indicators that reflect the core principles outlined in Section 4 of this framework. Indicators apply to recovery priorities and resource needs and set

realistic expectations and milestones for community members, stakeholders and supporting agencies.

- Ensure full community participation in developing metrics in coordination with local, State, Tribal and Federal partners. Include persons with disabilities and others with access and functional needs, individuals with limited English proficiency, seniors, members of underserved populations and advocates representing the unique needs of children.
- Leverage technology and systems innovations to achieve goals that result in greater information sharing, accountability and transparency.
- Assure that recovery activities respect the civil rights and civil liberties of all populations and do not result in discrimination on account of race, color, national origin (including limited English proficiency), religion, sex, age or disability.
- Ensure continuous improvement by evaluating the effectiveness of recovery activities.

Government agencies and private organizations that provide assistance are encouraged to have a system of tracking their coordination and assistance efforts, ensuring accountability and enabling prompt adjustments to meet ongoing and changing needs.

The suggested considerations listed below are also applicable for developing metrics.

BASELINE IMPACT ASSESSMENT

Provides a basis to define known community recovery issues to help understand the extent and dimensions of disaster impacts in order to chart a path to a realistic recovery end state.

DESIRED OUTCOME

Focuses on recovery impacts and overall results, not just a target number (e.g., number of families in permanent housing versus number of housing units constructed).

CROSS-SECTOR ASSESSMENT

Tracks progress across all sectors, including but not limited to, housing, environmental, business, employment, infrastructure, access to essential health and social services and overall community accessibility.

6. RECOVERY ROLES AND RESPONSIBILITIES

Successful recovery depends on all recovery stakeholders having a clear understanding of pre- and post-disaster roles and responsibilities (Figure 2). In keeping with the *National Disaster Recovery Framework (NDRF)* principles, clearly defined roles and responsibilities are a foundation for unity of effort among all recovery partners to jointly identify opportunities, foster partnerships and optimize resources.

Individuals and Households

Individuals and families need to plan and be prepared to sustain themselves in the immediate aftermath of a disaster. Those who prepare reduce personal stress, and they enhance their ability to undertake their own recovery and shape the future of their community's recovery.

The extent to which individuals and families adequately prepare for disasters has an impact on the success of the recovery. This includes carrying adequate insurance and maintaining essential levels of supplies, such as medication, food and water. Resources to help individuals and families prepare are available through websites and publications of various organizations that are active in disasters, including local, State, Tribal and Federal agencies. Maintaining awareness of public information on the recovery process helps to eliminate confusion and uncertainty.

Recommended roles and activities of individuals and families are detailed in Appendix B.

Partnerships at every level are supported by State and Federal authorities and encouraged through two-way communication.

Figure 2. Community-Focused Recovery.

Private Sector – Business Community and Critical Infrastructure Owners and Operators

The private sector plays a critical role in establishing public confidence immediately after a disaster. When the private sector is operational, the community recovers more quickly by retaining and providing jobs and a stable tax base. If local leadership and the business community work together pre-disaster and develop a conceptual recovery plan, the public is more likely to be optimistic about the community's ability to recover post-disaster.

Additionally, the private-sector owns and operates the vast majority of the Nation's critical infrastructure, such as electric power, financial and telecommunications systems. These entities play a major role in the recovery of a community or region as a whole.

It is critical that disaster recovery officials recognize the importance of partnership and create coordination opportunities during pre-disaster planning with private sector leaders. The resources and capabilities of the private-sector, including utilities, banks and insurance companies, can play an important role in encouraging mitigation and creating greater resilience in a community. For example, local banks can create products to encourage individuals and businesses to be financially prepared for disasters and work with small businesses to develop business continuity plans. Insurance companies can educate community members on risks, reach out to underserved populations and work with local, State and Tribal governments to find ways to provide coverage for families and businesses in the community.

Major players in recovery efforts, businesses and critical infrastructure owners and operators have an important responsibility to improve disaster resilience by mitigating risks and increasing disaster preparedness. Businesses should adopt and exercise business continuity plans to minimize costly operational disruptions and purchase adequate all-hazards insurance policies. Businesses that plan for disruption are less likely to go out of business after a disaster than those that do not.

Recommended private sector roles and activities are detailed in Appendix B.

Nonprofit Sector

The nonprofit sector plays a vital role in the recovery of impacted communities. Nonprofits include voluntary, faith-based and community organizations, charities, foundations and philanthropic groups as well as professional associations and academic institutions. The formidable value of the work of these stakeholders resides in community recovery planning, case management services, volunteer coordination, behavioral health and psychological and emotional support, technical and financial support, housing repair and construction that meets accessibility/universal design standards, and project implementation.

Nonprofit-sector support is provided by a range of organizations from small locally-based nonprofits to national organizations with extensive experience in disaster recovery. Nonprofits directly supplement and fill gaps where government authority and resources cannot be applied. Resourceful fundraisers, grantors and investors inject needed financial resources to meet recovery needs and obligations that otherwise are not funded by a government program.

Many organizations originate from or stay behind in the impacted community to continue to mobilize support and provide services. Particularly in a large-scale or catastrophic disaster, they play a critical role in the implementation of an inclusive, locally-led recovery organization and process during the transition as Federal and State recovery support recede and local leadership and community recovery organizations complete the mission.

Nonprofit organizations are critical for ensuring participation and inclusion of all members of the impacted community. Many nonprofits act as advocates for, or assistance providers to, a wide range of members of the community, such as individuals with disabilities and others with access and functional needs, children, seniors, individuals with limited English proficiency and other underserved populations. It is crucial that these individuals and families receive timely recovery information, participate in the recovery process and understand and have access to resources to achieve recovery.

Recommended nonprofit sector roles and activities are detailed in Appendix B.

Local Government

The local government has the primary role of planning and managing all aspects of the community's recovery. Individuals, families and businesses look to local governments to articulate their recovery needs. Those plans should include a *Continuity of Government* (*COG*) and *Continuity of Operations* (*COOP*) *Plan*[2]. Local government may become overwhelmed and need staffing, recovery expertise, leadership or other assistance. State and Federal officials work with local governments in the development and implementation of their plans and recovery efforts when needed and requested.

The majority of mitigation measures are adopted, codified and enforced at the local level. While there are State and Federal standards, it is often up to the local government to adopt and enforce them. Examples include participating in the *National Flood Insurance Program (NFIP)* and enforcing building codes.

Local governments also lead the community in preparing hazard mitigation and recovery plans, raising hazard awareness and educating the public of available tools and resources to enhance future resilience (e.g., Section 7 of the *Americans with Disabilities Act (ADA) Best Practices Tool Kit*, concerning emergency preparedness and people with disabilities). Government agencies play roles as employers and need their own plans to protect and assist employees during emergencies. Finding opportunities to share public information on the recovery process is important to maintaining community coordination and focus.

Recommended local government roles and activities are detailed in Appendix B.

State Government

States lead, manage and drive the overall recovery process and play the central role in coordinating recovery activities that include providing financial and technical support. States oversee regional coordination of recovery, set priorities and direct assistance where it is needed.

States are a conduit to local and Tribal governments for key Federal recovery assistance programs. In addition to managing Federally-provided resources, State government may develop programs or secure funding that can help finance and implement recovery projects. An example of this type of assistance is helping communities acquire appropriate insurance coverage pre-disaster or issuing bonds after a disaster. Where additional needs exist, States can reassign existing internal resources to streamline and expedite recovery, such as forming a new or ad hoc State recovery agency. States play an important role in keeping the public informed through strategic messaging and working with all other stakeholders to provide an information distribution process. State government agencies are also employers and need their own disaster recovery plan, such as *Continuity of Governments (COG)* and *Continuity of Operations (COOP)*, to protect and assist their employees.

Recommended roles and activities for State governments are detailed in Appendix B.

Tribal Government

Tribal governments, as sovereign nations, govern and manage the safety and security of their lands and community members. Many Tribal government borders cross multiple counties and States, presenting a unique challenge in planning response and recovery efforts. While resources in other communities and governments may be available and easily accessible, this is not the case in many Tribal government communities. Understanding these basic facts assists local, State and Federal governments when working with the sovereign Tribal governments to develop and implement their recovery plans.

The Federal Government is required to engage in meaningful consultation with Tribal governments prior to the finalization of policy or program implementation. Local and State governments are encouraged to engage with Tribal governments as well. (See *Executive Order 13175, Consultation and Coordination with Indian Tribal governments.*)

Recommended roles and activities for Tribal governments are detailed in Appendix B.

Federal Government

The Federal Government can play a significant facilitative role in the development of urban and rural communities and their social infrastructures, and can leverage needed resources to build and rehabilitate many communities so that they are more disaster resistant and resilient.

When a disaster occurs that exceeds the capacity of State and Tribal resources — or impacts Federal property, other areas of primary Federal jurisdiction or national security interests — the Federal Government may use the *National Disaster Recovery Framework (NDRF)* to engage necessary and available department and agency capabilities to support local recovery efforts.

The Federal Government's supporting role is especially important during the early weeks after a large-scale disaster or catastrophic incident, when many local, State and Tribal governments are overwhelmed with response and relief efforts. The duration and extent of Federal support is determined in part by the scale and enduring impacts of the disaster. The Federal Government's disaster recovery management and support systems must be scalable and adaptable so changes can be made quickly and effectively to meet the needs of the specific disaster.

The Federal Government also plays an important role in providing accessible information to the public and all stakeholders involved in recovery, including information about Federal grants and loans with potential applications to recovery. In coordination with local, State and Tribal communicators, the Federal Government is responsible for ensuring that information is distributed as well as understood, so that the public, Congress, the private-sector and all stakeholders are informed and aware of the process and realistic expectations for recovery.

Federal agencies may be directed by the President to provide Federal-to-Federal or other support. Federal agencies without recovery missions may directly or indirectly contribute to meeting recovery needs of affected communities by delivering assistance provided under their normal authority. The Federal Government coordinates its activities, programs and funding sources to facilitate adaptations and adjustments consistent with other competing requirements, including other disaster response and recovery needs.

Prior to a disaster, the Federal Government has a responsibility to assist local, State and Tribal governments to prepare for recovery by providing guidance and tools for planning and preparedness activities. Although disasters and localities vary so widely that most recovery planning must transpire at the local level, some centralized planning and Federal guidance or standards are necessary to ensure coordination of outside resources and assistance. Large-scale and catastrophic incidents (e.g., the Midwest Floods of 1993 and 2008, Hurricanes Katrina and Rita in 2005, the Gulf Coast Oil Spill of 2010, or a potential New Madrid Earthquake) often cross municipal, county, State or even Tribal jurisdictions. National coordination encourages unity of effort among government agencies and nongovernmental organizations (NGOs) to achieve the optimal benefit for those impacted.

From the Federal perspective, a successful recovery optimizes the return on Federal investment. This includes reducing future risk from hazards and increasing resilience while adopting courses of action consistent with national laws and policies. The Federal Government requires that all recipients of Federal assistance comply with civil rights obligations under Section 504 of the *Rehabilitation Act* and *Title VI of the Civil Rights Act of 1964*, and the *Age Discrimination Act of 1975*. Government agencies also play roles as employers and need to have their own plans to protect and assist their employees during emergencies.

Recommended roles and activities for the Federal Government are detailed in Appendix B. Roles and responsibilities of the Federal agencies that comprise the Recovery Support Functions (RSFs) are detailed in the RSF Annexes.

7. Leadership

Achieving Disaster Recovery (Section 5) describes the components of a successful disaster recovery management system for all levels of government decisionmaking. Coordination, integration, community engagement and management are prominent system elements in keeping with the *National Disaster Recovery Framework (NDRF)* Core Principles of Leadership. To lead these critical disaster recovery functions, this section describes:

- Recommended Recovery Manager and Recovery Coordinator positions at the local, State and Tribal levels.
- New designated Federal Disaster Recovery Coordinator (FDRC) positions.
- National-level disaster recovery coordination.

Local Disaster Recovery Managers, State and Tribal Disaster Recovery Coordinators

The *National Disaster Recovery Framework (NDRF)* strongly recommends that State governors as well as local government and Tribal leaders prepare as part of their disaster recovery plans to appoint Local Disaster Recovery Managers (LDRMs) and State/Tribal Disaster Recovery Coordinators (SDRCs/TDRCs) to lead disaster recovery activities for the jurisdiction.

The role of the LDRMs, SDRCs and TDRCs is to organize, coordinate and advance the recovery at the local, State or Tribal level. The experience and skill sets of these individuals should include a strong basis in community development and good knowledge of the community's demographics. While these positions will often interact with the emergency management community, it is not necessary that these individuals be emergency management professionals. Their primary role is to manage and coordinate the redevelopment and building of community. In addition, the individuals occupying the positions should be able to represent and speak on behalf of their respective chief executives (e.g., mayor, governor, Tribal leader). The LDRMs and TDRCs serve as the jurisdiction's primary point of contact (POC) with the SDRC.

In large-scale disasters and catastrophic incidents where a Federal role may be necessary, the SDRC and/or TDRC is the primary interface with the Federal Disaster Recovery Coordinator (FDRC). Depending on the severity of the incident and anticipated scope and duration of disaster recovery efforts, the State Coordinating Officer (SCO) may fulfill the Recovery Coordinator role under the *Robert T. Stafford Disaster Relief and Emergency Assistance Act (Stafford Act)*. However, after large-scale disasters or catastrophic incidents, States are encouraged to appoint a separate position to ensure recovery activities are well-managed while extended response and short-term recovery activities are ongoing.

Responsibilities of LDRMs and SDRCs/ TDRCs may include, but are not limited to those listed in Tables 1a, 1b and 1c on the preceding three pages. In some cases, these responsibilities overlap, as shown.

Federal Disaster Recovery Coordinator

While disaster-impacted jurisdictions must necessarily and immediately focus on emergency response activities, the decisions made very early after a disaster influence recovery. In large-scale disasters and catastrophic incidents where a Federal role may be necessary, the Federal Disaster Recovery Coordinator (FDRC) is a focal point for incorporating recovery and mitigation considerations into the early decisionmaking processes.

The FDRC monitors the impacts and results of such decisions and evaluates the need for additional assistance and adjustments where necessary and feasible throughout the recovery.

In these situations, the FDRC works as a deputy to the Federal Coordinating Officer (FCO) for all matters concerning disaster recovery. The FDRC is responsible for facilitating disaster recovery coordination and collaboration between the Federal, Tribal, State and local governments, the private sector and voluntary, faith-based and community organizations. The FDRC partners with and supports the Local Disaster Recovery Manager (LDRM) and the State and/or Tribal Disaster Recovery Coordinator (SDRC/TDRC) to facilitate disaster recovery in the impacted State or Tribal area.

Table 1a. Local Disaster Recovery Manager (LDRM)

Pre-Disaster Responsibilities
1) Serve as primary point of contact (POC) for disaster recovery preparedness with the State and neighboring local and Tribal governments. 2) Coordinate development, training and exercise of jurisdiction disaster recovery plan (Section 9). 3) Establish and maintain contacts and networks for disaster recovery resources and support systems. 4) Promulgate principles and practices that further resiliency and sustainability in development and strategic planning initiatives.
Post-Disaster Responsibilities
1) Lead the creation and coordinate the activities of local recovery-dedicated organizations and initiatives. 2) Work with the SDRC to develop a unified and accessible communication strategy. 3) Participate in damage and impact assessments with other recovery partners. 4) Organize recovery planning processes, which includes individuals with disabilities and others with access issues, seniors and members of underserved communities, to fully engage constituents' input; lead the development of the community's or Tribe's recovery visions, priorities, resources, capability and capacity. 5) Ensure inclusiveness in the community recovery process, including persons with disabilities and limited English proficiency. 6) Communicate recovery priorities to State and Federal governments and other recovery stakeholders and supporters. 7) Incorporate critical mitigation, resilience, sustainability and accessibility-building measures into the recovery plans and efforts. 8) Lead the development of the community's recovery plans and ensure that they are publicly supported, actionable and feasible based on available funding and capacity. 9) Collaborate with State, Federal and other stakeholders and supporters, such as the business and nonprofit communities, to raise financial support (including long-term capital investment in local businesses) for the community's recovery, leverage the resources where possible and resolve potential duplication of assistance. 10) Work closely with the recovery leadership at all levels to ensure a well-coordinated, timely and well-executed recovery. 11) Develop and implement recovery progress measures and communicate adjustments and improvements to applicable stakeholders and authorities.

Table 1b. Tribal Disaster Recovery Coordinator (TDRC)

Pre-Disaster Responsibilities
1) Serve as primary POC for disaster recovery preparedness with neighboring local and Tribal governments, the Federal Government, and the State(s). 2) Coordinate development, training and exercise of jurisdiction disaster recovery plan (Section 9). 3) Establish and maintain contacts and networks for disaster recovery resources and support systems. 4) Promulgate principles and practices that further resiliency and sustainability in development and strategic planning initiatives.
Post-Disaster Responsibilities
1) Lead the creation and coordinate the activities of Tribal recovery-dedicated organizations and initiatives. 2) Work with the SDRC to develop a unified and accessible communications strategy. 3) Work with recovery coordinators and leads at the Federal and other levels to facilitate the development of a unified and accessible communications strategy. 4) Participate in damage and impact assessments with other recovery partners. S) Organize and support recovery planning processes, which includes individuals with disabilities and others with access issues, seniors and members of underserved communities, to fully engage constituents' input and result in development of the Tribe's recovery visions, priorities, resources, capability and capacity. 6) Ensure inclusiveness in the community recovery process, including persons with disabilities and limited English proficiency. 7) Communicate recovery priorities to State and Federal governments and other recovery stakeholders and supporters. 8) Incorporate critical mitigation, resilience, sustainability and accessibility-building measures into the recovery plans and efforts. 9) Lead the development of the Tribe's recovery plans and ensure that they are publicly supported, actionable and feasible based on available funding and capacity. 10) Coordinate State, Tribal, Federal and other funding streams for recovery efforts and communicate issues and solutions to recovery assistance gaps and overlaps. 11) Collaborate with State, Federal and other stakeholders and supporters, such as the business and nonprofit communities, to raise financial support (including long-term capital investment in local businesses) for the Tribe's recovery, leverage the resources where possible and resolve potential duplication of assistance. 12) Work closely with the recovery leadership at all levels to ensure a well-coordinated, timely and well-executed recovery. 13) Develop and implement recovery progress measures and communicate adjustments and improvements to applicable stakeholders and authorities.

FDRC Authority

Nothing in the *National Disaster Recovery Framework (NDRF)* alters or impedes the ability of local, State, Tribal or Federal departments and agencies to carry out their specific authorities or perform their responsibilities under all applicable laws, Executive Orders and directives. FDRC authority to facilitate disaster recovery coordination and collaboration is derived from the appropriate disaster recovery authority that may apply to the incident.

Other Federal departments and agencies carry out their disaster recovery authorities and responsibilities within the overarching construct of the NDRF. Additionally, nothing in the NDRF is intended to impact or impede the ability of any Federal department or agency to take an issue of concern directly to the President or any member of the President's staff [3]. For a large-scale disaster or catastrophic incident declared under the *Robert T. Stafford Disaster*

Relief and Emergency Assistance Act (Stafford Act), the FDRC works as a deputy to the FCO for all matters concerning disaster recovery.

FDRC Qualifications, Selection and Credentialing

FDRCs are selected based on their knowledge and experience with disaster recovery, mitigation, community development, resiliency planning, public administration concepts, and the range of Federal programs and interagency processes required for effective implementation of recovery initiatives. In addition to existing subject- matter expertise, FDRCs maintain training and credentialing emphasizing consistent and effective practices. Qualified FDRCs are senior level officials empowered to directly access designated senior officials in every Federal agency that may contribute to recovery. FEMA will appoint the FDRC.

Table lc. State Disaster Recovery Coordinator (SDRC)

Pre-Disaster Responsibilities
1) Serve as primary POC for disaster recovery preparedness with local, Tribal and Federal governments, particularly the Federal Disaster Recovery Coordinator (FDRC). 2) Coordinate development, training and exercise of jurisdiction disaster recovery plan (Section 9). 3) Establish and maintain contacts and networks for disaster recovery resources and support systems. 4) Promulgate principles and practices that further resiliency and sustainability in development and strategic planning initiatives.
Post-Disaster Responsibilities
1) Establish and/or lead a statewide structure for managing recovery. Provide support for local and/or Tribal recovery-dedicated organizations initiatives. 2) Communicate the roles and responsibilities of the State to the local and Tribal governments. 3) Work with recovery coordinators and leads at the Federal and other levels to facilitate the development of a unified and accessible communication strategy. 4) Support recovery planning processes, which includes individuals with disabilities and others with access issues, seniors and members of underserved communities, to fully engage constituents' input and result in development of the community's or Tribe's recovery visions, priorities, resources, capability and capacity. 5) Ensure inclusiveness in the community recovery process, including persons with disabilities and limited English proficiency. 6) Facilitate communication of recovery priorities for all impacted communities. Communicate statewide recovery priorities to the FDRC. 7) Encourage incorporation of critical mitigation, resilience, sustainability and accessibility-building measures into the recovery plans and efforts. 8) Coordinate State, Tribal, Federal and other funding streams for recovery efforts and communicate issues and solutions to recovery assistance gaps and overlaps. 9) Collaborate with Federal and other stakeholders and supporters, such as the business and nonprofit communities, to raise financial support (including long-term capital investment in local businesses) for the community's or Tribe's recovery, leverage the resources where possible and resolve potential duplication of assistance. 10) Work closely with the recovery leadership at all levels to ensure a well-coordinated, timely and well-executed recovery. 11) Develop and implement recovery progress measures and communicate adjustments and improvements to applicable stakeholders and authorities.

FDRC Pre-Disaster Regional Engagement

The responsibilities of the FDRC require an understanding of pre-disaster recovery planning as well as post-disaster recovery leadership and coordination. Since each community is unique in terms of its size, population and challenges, the development of effective recovery efforts will need to be crafted to fit each individual region's risks and needs. Therefore, it will be extremely helpful to the success of an FDRC to have pre-established relationships with persons at the Federal, Tribal, State and local levels, including the private and nonprofit sectors. In large-scale and catastrophic incidents where a Federal role may be necessary, the FDRCs have the knowledge, connections, and relationships to immediately begin effective disaster recovery coordination.

FDRC Post-Disaster Responsibilities

In large-scale disasters and catastrophic incidents when it may be necessary to deploy an FDRC in partnership with the State the FDRC's post-disaster responsibilities may include:

- **Develop a strategic approach for coordinating Federal assistance and policies.** The intent is to facilitate timely, sufficient and effective Federal assistance to the impacted State or Tribal government to support its disaster recovery.
- **Work with the impacted community to establish relevant recovery measures.** The aim is to track overall recovery progress and support the community in meeting its recovery goals in terms of outcome, milestones and budget; to make timely adjustments to the recovery effort if needed; and to define relationships between new players and the existing framework.
- **Promote inclusiveness in recovery.** The intent is to increase participation of stakeholders to ensure innovations and solutions that support recovery are considered. The community should provide a forum to engage disaster- impacted individuals, particularly individuals with disabilities, individuals with limited English proficiency, seniors, members of underserved populations and advocates for children so that their needs and contributions are an integral part of the recovery process and outcome.
- **Facilitate the development of a unified communications strategy.** The objective is to have all stakeholders work in concert to manage expectations and to communicate a clear, consistent message to the public and ensure an accessible, comprehensive and culturally and linguistically appropriate communications outreach strategy.
- **Coordinate Federal assistance to support community recovery planning.** The goal is to supplement local capacity with needed expertise to conduct a successful planning process that results in a recovery plan that is publicly supported, actionable and leverages available resources.
- **Work with the impacted community to incorporate mitigation and resilience-building measures into recovery plans and implementation.** The goal is to minimize the community's risk to all hazards and make the recovered community safer, stronger, sustainable and more resilient from any man-made or natural hazards.
- **Coordinate the Recovery Support Function (RSF) operations and activities.** The FDRC consults with the RSF field coordinators to conduct a recovery impact assessment and recommend activation of the appropriate RSFs. The objective is to

focus Federal resources on the most pertinent recovery needs and to promote partnerships between the Federal Government and stakeholders at the local, State and Tribal levels.

- **Facilitate Federal funding streams and solutions to assistance gaps and overlaps.** The intent is to maximize the benefit from Federal funds that an impacted community is qualified to receive, help prevent recovery delays, resolve rule and regulatory conflicts to the extent possible and help eliminate possible duplication of assistance in coordination with local, State and Tribal recovery coordinators.
- **Reinforce the importance of compliance with Federal civil rights laws when using Federal funds.** Federal funding carries with it the responsibility to comply with anti-discrimination laws. Federally-funded programs and activities should not intentionally or unintentionally exclude groups of people as a result of race, color, national origin, limited English proficiency, religion, sex, age or disability.

Annexes, Standard Operating Procedures (SOPs) and other supporting tools and documents will provide additional guidance relating to FDRC post-disaster responsibilities.

Activation, Transition and Demobilization

Activation, transition and demobilization of the FDRC and Recovery Support Functions (RSFs) depends on the magnitude of the disaster, requirements of affected communities, and availability and appropriateness of Federal resources. During large-scale and catastrophic incidents, the FDRC may be deployed to serve as Deputy to the FCO and primary advisor to the FCO on all recovery issues. The FDRC supports the FCO and coordinates Federal recovery operations on his or her behalf.

The NDRF employs an assessment protocol to ensure a scalable, flexible, adaptable and cost- effective approach to recovery activities and to determine which coordination structures are necessary and appropriate under the circumstances. From this assessment, the FCO, in coordination with the State, activates the appropriate Recovery Support Functions (RSFs), if necessary.

The FDRC coordinates with the FCO and the NDRP at FEMA Headquarters on the progress of recovery, including identifying policy and program challenges. The FDRC may also be called upon to brief senior level officials in the Executive Branch and Congress on the pace, challenges and needs of the recovery, and to propose and coordinate solutions.

The FDRC leads the Federal Government's effort to develop a *Recovery Support Strategy* that supports the recovery needs of impacted communities, State and Tribes.

The *Recovery Support Strategy* articulates how all Federal agencies participate in the recovery and coordinate support based on the needs identified through the impact assessment process and local, State and Tribal recovery plans. It includes a comprehensive timeline, with key milestones and benchmarks that guide the ongoing operations and eventual demobilization of the FDRC and other Federal resources in the impacted area. Throughout the recovery, the FDRC facilitates the coordination of information and activities among the Federal agencies whose programs, technical assistance and expertise are relevant to recovery, within the framework of the *Recovery Support Strategy*.

Transition. The coordination between ESFs and RSFs is the responsibility of the FCO/RSF/ FDRC team. Disaster operations vary based on the nature, scope and complexity of the specific incident. Therefore, the timing of the transition from the response to initial recovery operations and then to recovery varies. During response and in the early stages of recovery, RSFs may be deployed while ESFs are still operational and the two coexist until the ESFs fully demobilize. Working together in collaboration with Tribal, State and local authorities, the FCO determines when it is appropriate to begin phasing out the Emergency Support Function (ESF) and Joint Field Office (JFO) elements associated with the *National Response Framework (NRF)*.

In large-scale disasters and catastrophic incidents, the FDRC takes over the lead from the FCO, when the FCO demobilizes, to continue management of Federal recovery resources, for those incidents that require continued significant interagency disaster recovery coordination. This includes coordination of the longer-term RSF structures associated with the NDRF that continue operation.

Transition involves a conscious effort, from day one of the recovery operation, to actively engage and encourage local, State and Tribal leadership and ownership of the recovery process. It provides coordination support and technical assistance, with the intent to supplement, not substitute, local leadership, ownership and capabilities.

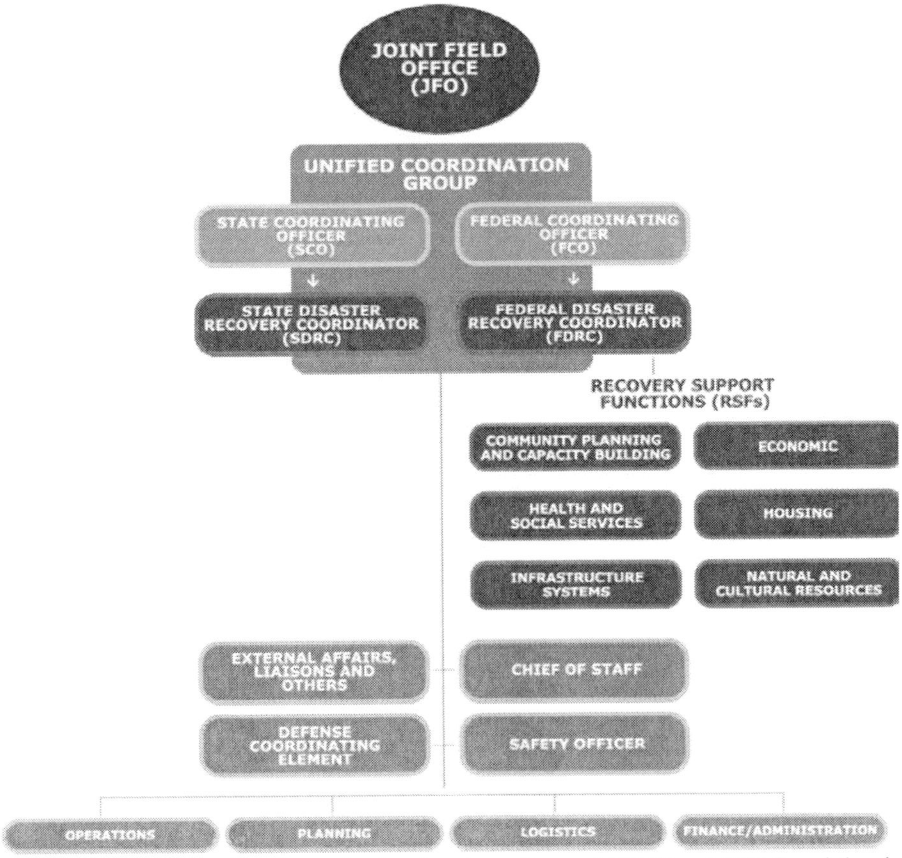

JFO organizational structure for the newly developed positions of FDRC and SDRC and the six RSFs established within the NDRF.

Figure 3. Recovery Functions (FDRC, SDRC and RSFs) within the Joint Field Office Chain of Command.

Using the *Recovery Support Strategy* as a guide, the FDRC continuously tracks the recovery progress, evaluates the adequacy and pace of recovery assistance and works with local, State, Tribal, Federal, nonprofit, faith- based and private-sector stakeholders to identify gaps and/or additional support needs. The FDRC is responsible for closely coordinating this progress, tracking timelines and communicating information with local, State and Tribal officials as well as other key stakeholders. Close communication and coordination with the local, State and Tribal governments and stakeholders throughout the recovery process reinforces a shared understanding of the objectives and expectations for the unified disaster recovery effort and eventual Federal demobilization.

Demobilization. How long the FDRC remains on-site in the disaster area depends upon the scale of the disaster and on-site coordination requirements. Regardless of the length of sustained on-site presence, the FDRC may remain closely engaged with local and State officials for an extended period. When the impacted local, State or Tribal government has recovered the capacities and resources needed to manage its disaster recovery effort, the FDRC has already commenced — if not nearly completed — the transition of its recovery role and responsibility to the LDRMs and SDRCs/ TDRCs. This transition involves regional Federal staff assuming greater roles in coordinating ongoing Federal support. Once this transition and coordination with the impacted State and/or Tribal government is complete the FDRC demobilizes from the mission and exits the impacted area.

Disaster Recovery Coordination

Disaster recovery coordination occurs through partnerships with the Recovery Support Function (RSF) agencies. The National Disaster Recovery Planning (NDRP) Division within the Recovery Directorate of the Office of Response and Recovery at FEMA Headquarters serves as the focal point for all interagency coordination for disaster recovery issues at the national level and is responsible for the ongoing implementation of the *National Disaster Recovery Framework (NDRF)*. The NDRP Division will also facilitate regional coordination among Federal Disaster Recovery Coordinators (FDRCs) following a large-scale disaster or catastrophic incident that requires significant interagency recovery resource coordination in multiple States. The NDRP Division ensures that resiliency, mitigation, inclusiveness and other central concepts of the NDRF are appropriately addressed in disaster recovery operations. An important function of this component is to improve coordination and delivery of Federal programs that assist with disaster recovery, increase local and State disaster recovery management capacity and improve disaster resiliency nationwide.

Pre-Disaster Role

The NDRP Division oversees the designation, training, credentialing, deployment and evaluation of Federal Disaster Recovery Coordinators (FDRCs). It also coordinates and supports the efforts of the national RSF coordinating agencies (Section 8) to develop annexes, Standard Operating Procedures (SOPs) and other supporting tools and documents.

The NDRP Division, in close collaboration with the RSF agencies, coordinates Federal guidance and training to assist local, State and Tribal governments with disaster recovery preparedness. This includes planning, organizational development and management capacity building, building community resilience, training, exercise, evaluation and improvement. In addition, the Division captures, manages and ensures sharing of a repository of disaster recovery best practices, lessons learned and other data.

The NDRP Division also is charged with convening RSF coordination meetings, as necessary, to discuss ongoing recovery operations and agency efforts to promulgate resiliency into steady-state programs and policies.

Post-Disaster Role

The NDRP Division is a primary Federal focal point for disaster recovery support. After an incident, the NDRP Division may provide technical assistance to the Federal Coordinating Officer (FCO) to determine if a FDRC deployment is appropriate. During open disaster recovery operations, the Division provides consultation support and facilitates coordination with executive level leadership for the FDRC and deployed RSFs. It also coordinates the efforts of the RSF coordinating agencies at the national level to support their field components. In large-scale disasters and catastrophic incidents, the Division supports the deployed FDRC to coordinate recovery efforts and the deployed RSFs

8. Recovery Support Functions

The Recovery Support Functions (RSFs) comprise the *National Disaster Recovery Framework's (NDRF's)* coordinating structure for key functional areas of assistance. Their purpose is to support local governments by facilitating problem solving, improving access to resources and by fostering coordination among State and Federal agencies, nongovernmental partners and stakeholders.

The RSFs created within the NDRF bring together the core recovery capabilities of Federal departments and agencies and other supporting organizations — including those not active in emergency response — to focus on community recovery needs. The RSFs are organized into six manageable components and through the RSFs, relevant stakeholders and experts are brought together during steady- state planning and when activated post-disaster to identify and resolve recovery challenges. RSFs and stakeholders organize and request assistance and/or contribute resources and solutions. Together, these RSFs help facilitate local stakeholder participation and promote intergovernmental and public-private partnerships.

Recovery Support Functions

Community Planning and Capacity Building.
Economic.
Health and Social Services.
Housing.
Infrastructure Systems.
Natural and Cultural Resources.

Differences between ESFs and RSFs

The Recovery Support Function (RSF) structure coexists with and builds upon the Emergency Support Functions (ESFs) under the *National Response Framework (NRF)*. RSFs are different from ESFs in that they have different mission objectives, partnerships, approaches, time spans and organizational structure; additionally, the players and skill sets involved may be different.

Mission Objectives

The objective of the RSFs is to facilitate the identification, coordination and delivery of Federal assistance needed to supplement recovery resources and efforts by local, State and Tribal governments, as well as private and nonprofit sectors. An additional objective is to encourage and complement investments and contributions by the business community, individuals and voluntary, faith-based and community organizations. These RSF activities assist communities with accelerating the process of recovery, redevelopment and revitalization.

Players and Skill Sets

RSF staff may require different skill sets than their colleagues from the same agencies working under the ESF structure. For example, the skills needed to provide disaster sheltering services under ESF #6 are different than those necessary to address long-term housing solutions supported by the Housing RSF.

Partnerships

RSFs involve partners in the local, State and Tribal governments and private and nonprofit sectors not typically involved in emergency support functions but critically needed in disaster recovery. These new partners may include public and private organizations that have experience with permanent housing financing, economic development, advocacy for underserved populations and long-term community planning.

Approaches

The processes used for facilitating recovery are more flexible, context based and collaborative in approach than the task-oriented approach used during the response phase of an incident. Recovery processes should be scalable and based on demonstrated recovery needs.

Time Spans

Whereas the ESFs typically operate within a time span of weeks and months, the RSF operational timeframe is months to years. RSFs will likely activate before all ESFs demobilize; therefore they may coexist within the same operation for a period of time. Neither ESFs nor RSFs have a predetermined point at which they demobilize.

Organizational Structure

The RSFs are:

- **Community Planning and Capacity Building**
- **Economic**
- **Health and Social Services**
- **Housing**
- **Infrastructure Systems**
- **Natural and Cultural Resources**

Relationship between ESFs and RSFs

Recovery cannot wait until those occupied with response and short-term recovery activities have time and space to start thinking about recovery. A discrete and well-resourced recovery focus, operating at the same time as response activities, is established to ensure that communities transitioning out of response are positioned to find themselves ahead of the curve in organizing and planning for major reconstruction and redevelopment necessary for recovery.

As the level of response activities declines and recovery activities accelerate, the Federal Disaster Recovery Coordinator (FDRC) will engage with the Recovery Support Function (RSF) agencies to organize and coordinate Federal recovery assistance. During this early recovery phase, the FDRC and the RSF coordinators are working closely with Emergency Support Function (ESF) leads to share information about impacts, assistance provided and working relationships at all levels. There is necessarily some overlap between the ESF and RSF missions, but as the ESF requirements diminish, and the recovery issues take center stage, the RSFs take over the residual ESF activities that are associated with recovery. The timing of the transition from ESF to RSF depends on the nature of the activity, and may vary considerably from RSF to RSF. The Federal Coordinating Officer (FCO) determines when a specific ESF is no longer required.

It is essential to the success of the *National Disaster Recovery Framework (NDRF)* that Federal partners address responsibilities across the recovery continuum, including preparedness, mitigation and development activities as well as post-incident stabilization and recovery actions. The coordinator for each RSF creates detailed supporting guidance and tools for recovery implementation. The development of these RSFs is an iterative process that includes addressing gaps in authorities and resources.

RSF Roles and Responsibilities

Each Recovery Support Function (RSF) has a designated coordinating agency along with primary agencies and supporting organizations with programs relevant to the functional area. The RSF coordinating agency, with the assistance of the Federal Emergency Management Agency (FEMA), provides leadership, coordination and oversight for that particular RSF. Throughout the preparedness, response and recovery phases, the coordinating agency ensures ongoing communication and coordination between the primary agencies and support organizations, and between the Federal agencies and corresponding local, State and Tribal authorities and nonprofit and private sector organizations.

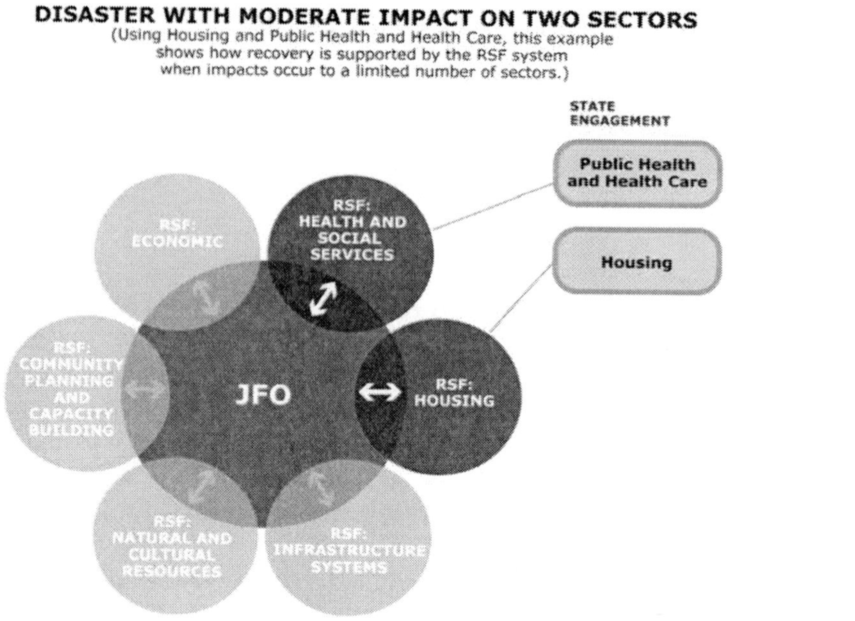

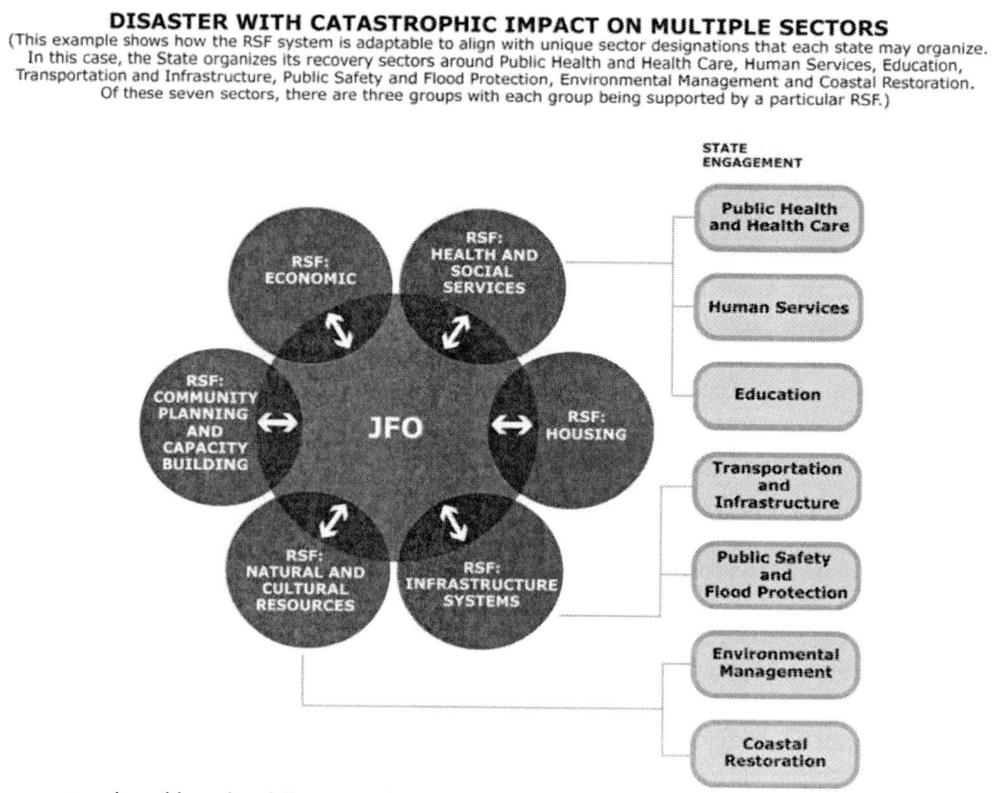

RSF support varies with scale of disaster and sectors impacted.

Figure 4. Scenarios of Different Levels of RSF Assistance.

An RSF primary agency is a Federal agency with significant authorities, roles, resources or capabilities for a particular function within an RSF. Primary agencies orchestrate Federal support within their functional area for an affected State and may lead interagency field assessment or support teams as necessary. Support organizations are those entities with specific capabilities or resources that support the primary agency in executing the mission of the RSF. The principal distinction between a primary and a supporting agency is the frequency with which the agency may be expected to actively participate in a RSF operation. RSF agencies provide assistance when requested by the Federal Disaster Recovery Coordinator (FDRC) or the designated RSF coordinator, consistent with their authority and resources, or as directed pursuant to section 402 of the *Robert T. Stafford Disaster Relief and Emergency Assistance Act (Stafford Act).*

When coordinating agencies are activated to lead an RSF, primary agencies and supporting organizations are expected to be responsive to RSF-related communication and coordination needs.

NDRP Division and FDRC

The National Disaster Recovery Planning (NDRP) Division at FEMA Headquarters serves as a focal point for all interagency coordination for disaster recovery issues at the headquarters level. The NDRP facilitates and coordinates Recovery Support Function (RSF) activities at the national level through the designated RSF coordinating and primary agencies.

During post-disaster operations, the RSF coordinating agencies, through an appropriate funding vehicle, report to the Federal Disaster Recovery Coordinator (FDRC) and lead their respective RSF members to facilitate the identification, coordination and delivery of Federal assistance needed to supplement State and local recovery resources. RSF staff deployed to the field, through an appropriate funding vehicle, report to and are coordinated by the FDRC assigned to the Joint Field Office (JFO).

Scalability and Adaptability

The Recovery Support Function (RSF) coordinating structure is scalable and adaptable to meet different levels and types of needs, as well as specific recovery requirements of large to catastrophic incidents (Figures 4 and 5). Each of the six RSFs has a pre-designated coordinating agency that works with the Federal Disaster Recovery Coordinator (FDRC) to promote communication and collaboration among its members. This tiered leadership structure helps to accommodate the rapid surge of Federal resources that may be needed to assist in large-scale or catastrophic incidents. Furthermore, through the RSFs, Federal resources are organized into a number of field teams led by the most appropriate primary agencies to cover multiple localities. Each team is then adapted to comprise only the RSF functions (or the Federal department or agency) that have the authority, expertise and resources appropriate to the locality assigned. Based on assessments and recovery management structures established by State and local officials, only the RSFs that are needed deploy.

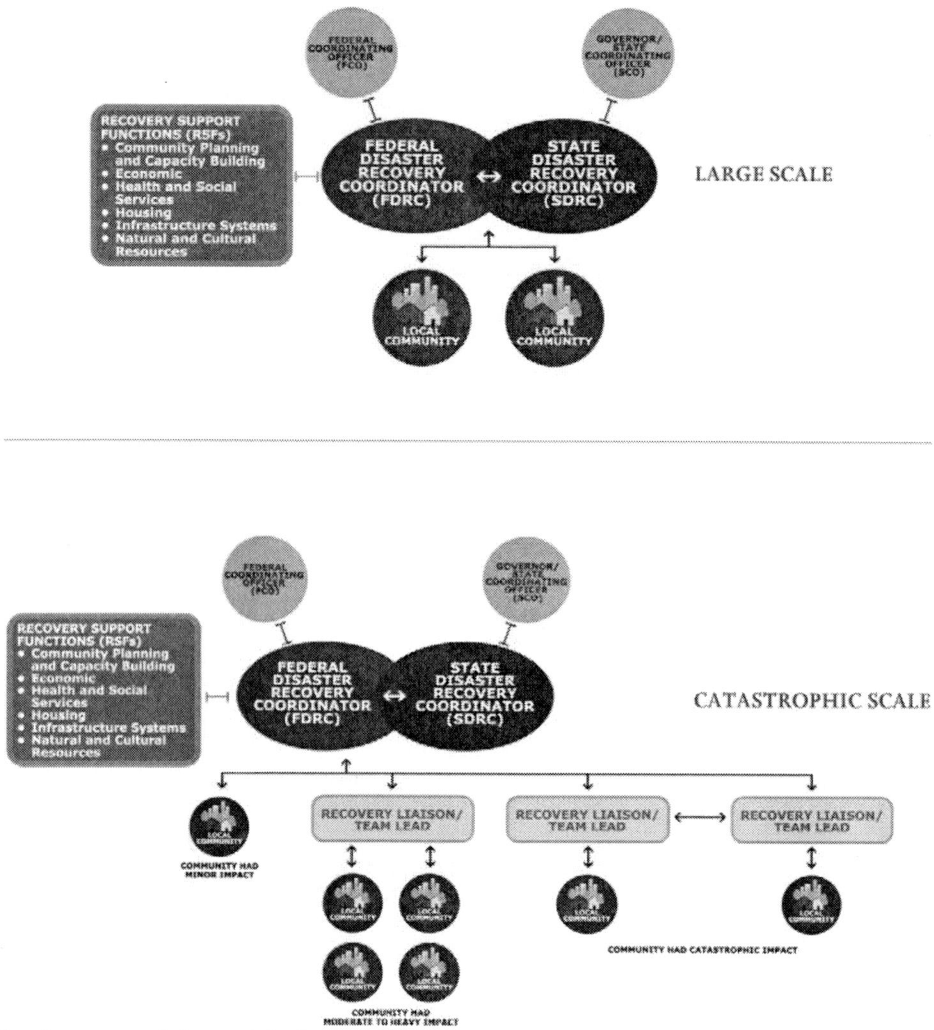

Levels of support and the need for Recovery Liaisons or Team
Leads vary depending on size and scale of a disaster.

Figure 5. Coordinating Structure for Large-Scale and Catastrophic Events.

Partnership and Inclusiveness

The Federal Government uses an inclusive process to ensure coordination with local and State elected officials to identify priorities for the application of Federal resources. In engaging with disaster-affected communities, the Recovery Support Functions (RSFs) seek to specifically include and address the needs of individuals with disabilities, those with access and functional needs, children, seniors, individuals with limited English proficiency and members of underserved populations.

The RSFs work closely with local, State and Tribal governments to identify underserved populations. The RSFs also coordinate with Federal Tribal Liaisons (FTR), Voluntary Agency Liaisons (VALs), Disability Issue Advisors (DIA) and other Federal offices, bureaus and programs when necessary. Local nongovernmental organizations (N GOs) and community groups often have excellent relationships with the underserved populations. The Federal Disaster Recovery Coordinators (FDRCs), through the RSFs, collaborate with these organizations to ensure that programs are culturally appropriate and that at-risk populations and their needs are identified.

In all actions, FDRCs and RSFs strive for affected residents to have a voice; for services to reach those who need them most; for equitable distribution of resources; and for recovery programs appropriate for the socioeconomic and cultural makeup of the community.

Resource Information

Each of the Recovery Support Function (RSF) member agencies brings a wealth of expertise and programmatic authorities and resources to the table. Annexes to the *National Disaster Recovery Framework (NDRF)* for each RSF outline in more detail how RSFs contribute to increases in community disaster resilience by coordinating to ensure cost-effective and efficient delivery of assistance. The annexes will also define how risk information and risk reduction technical expertise will be integrated into the work of the RSF in support of community recovery, including promotion of the use of the most appropriate and cost-effective building materials during rebuilding.

In addition, the *National Disaster Housing Strategy*, published in 2009, led to the creation of a web-based tool that collects data on different programs to help communities rebuild after a disaster. This web-based tool, the National Disaster Recovery Program Database (NDRPD), is a central location for local, State and Tribal governments and emergency managers to view recovery programs from governments, for-profit, nonprofit, and charitable organizations. The NDRPD provides easy access to information on programs, improves the visibility of programs that can help communities and enables communities to focus on those programs that can best suit their needs. The NDRPD may be accessed at https://asd.fema.gov/inter/ndhpd/public/home.htm.

Although the program information on the NDRPD is available for anyone to view, it is designed for local, State and Tribal governments and emergency managers. The database's programs are not intended for individual applicants; individuals and families may visit http://www.disasterassistance.gov to find and apply for individual assistance related to disaster recovery.

Overview of RSF Mission and Objectives

The mission and objectives of each of the six Recovery Support Functions (RSFs) are specified in the following pages. Once authorized and funded, the agencies participating in each RSF collaboratively develop operational guidance for use in recovery preparedness and disaster recovery operations. When working under an appropriate funding mechanism, RSFs operate under the leadership of the Federal Disaster Recovery Coordinator (FDRC). The FDRC utilizes overarching coordinating constructs to effectively manage and consolidate the RSF outputs when there is a large-scale or catastrophic incident. Their purpose is to support local governments by facilitating problem solving, improving access to resources, integrating principles of resiliency, sustainability and mitigation and fostering coordination among State and Federal agencies, nongovernmental partners and stakeholders.

RSFs develop guidance and standard operating procedures for rapid activation of their capabilities to support community recovery. Each RSF identifies relevant statutory and/or regulatory programs, potential capabilities and/or limiting factors pertaining to recovery support for their functional area of assistance. RSFs provide a forum for interagency coordination, information sharing and exchange of effective practices. RSFs may also support planning, preparedness, education, training and outreach efforts to enhance capabilities for recovery. Each RSF works with partners to identify critical facilities and ensure considerations are made to reduce risk pre- and post-disaster.

The full name of the coordinating, primary agencies and support organizations are found in Section 11, Abbreviations. Their common acronyms are used in the descriptions of the RSFs for brevity.

The following tables outline the key aspects of the six RSFs. They identify the coordinating and primary agencies as well as the supporting organizations associated with each RSF. The mission and function of each RSF is also explained. Key pre- and post-disaster activities as well as expected outcomes are highlighted for each RSF.

Table 2. Pre-Disaster, the Community Planning and Capacity Building RSF

- Coordinates the provision of preparedness planning and technical assistance support to aid Tribes, States and local governments to develop effective pre-disaster recovery plans that guide the full range of recovery efforts, both short- and long-term, and ensure all affected populations are included.
- Coordinates the resolution of outstanding Federal agency program and policy issues identified in after-action and other evaluations that present ongoing barriers or challenges for effective support for State, Tribal and local community planning and capacity necessary to facilitate an effective recovery process.
- Develops multidisciplinary recovery tools and best practices.
- Promotes resiliency measures and enhances coordination of programs that build local leadership capacity, community member involvement, partnerships and education on disaster preparedness for recovery.
- Promotes the importance of pre-disaster mitigation as an essential component of pre- disaster community recovery preparedness planning, including use of multihazard risk assessment.
- Identifies and leverages programs that assist communities to prepare, collect and analyze relevant existing and future data necessary to plan and manage complex disaster recovery.
- Integrates mitigation, recovery and other pre-disaster plans and activities into existing local, State and Tribal community-wide planning and development activitites, such as comprehensive plans, land use plans, economic development plans, affordable housing plans, zoning ordinances and other development regulations through technical assistance.
- Coordinates educational and cross-training opportunities for key participants in community recovery planning and capacity support including, but not limited to: emergency managers; city managers; planning, economic development and other local officials; and nonprofit and private sector partners for recovery.
- Develops pre-disaster partnerships with others such as Federal agency extension programs, universities, national professional associations, and nongovernmental organizations, to facilitate recovery capacity-building activities and expansion of resources available to communities after a disaster for planning and decisionmaking

Table 3. Post-Disaster, the Community Planning and Capacity Building RSF

- Maintains robust and accessible communications throughout the recovery process between the Federal Government and all other partners to ensure ongoing dialogue and information sharing.
- Identifies the range and significance of the disaster's effects on Tribes, regions and local governments in the impacted area.
- Coordinates the provision of resources to units of government for recovery planning technical assistance and to support recovery capacity and surge needs in a variety of Tribal/county/city functional areas (e.g., city Management, financial management, hazard mitigation and risk assessment, damage assessment, building inspection and permitting); coordinates resources to address other skill sets that communities often lack capacity after large-scale and catastrophic disasters.
- Develops community-focused technical assistance teams for uniquely or heavily impacted Tribes or communities, integrating the use of Federal agency resources organized under other RSFs.
- When activated by the FDRC, the primary and supporting departments and agencies deploy in support of the Community Planning and Capacity Building mission.
- Identifies and tracks resolution of gaps and conflicts in multiple Federal planning requirements and assistance programs, as well as programs that support and build community capacity and surge needs for recovery management.
- Coordinates the application and treatment of hazard mitigation and sustainability principles in Federally supported recovery planning efforts.
- Coordinates Community Planning and Capacity Building supported community-centric technical assistance teams with the establishment of local unmet needs committees or groups for assisting individuals and families.
- Aids local, State and Tribal governments to identify and integrate the consideration of all affected stakeholders, including vulnerable populations and persons with disabilities, and individuals with limited English proficiency into the public sector recovery plans and decisionmaking process.
- Provides technical assistance and planning support to aid all levels of government to integrate sustainability principles, such as adaptive re-use of historic properties, mitigation considerations, smart growth principles and sound land use into recovery decisionmaking and planning during the post-disaster period.
- Captures after-action recommendations and lessons learned.

RSF: Community Planning and Capacity Building

Coordinating Agency: DHS/FEMA.
Primary Agencies: DHS/FEMA, HHS.
Supporting Organizations: CNCS, DHS, DOC, DOI, DOJ, DOT, ED, EPA, GSA, HUD, SBA, TREAS, USDA.

Mission

Supporting and building recovery capacities and community planning resources of local, State and Tribal governments needed to effectively plan for, manage and implement disaster recovery activities in large, unique or catastrophic incidents.

Table 4. Outcomes for the Community Planning and Capacity Building RSF

Through a coordinated effort that draws from resources of Federal departments, agencies and services, the Community Planning and Capacity Building RSF provides expertise to ensure: • Enhanced interagency coordination of resources, requirements and support for building community capacity and community recovery planning. • Increased community self-reliance and adaptability. • Hazard mitigation and risk reduction opportunities have been integrated into all major decisions and reinvestments during the recovery process. • An improved planning process that ensures a more effective and efficient use of Federal, State, nongovernmental and private sector funds. • Communities are able to shorten the timeline and improve specific recovery outcomes through more effective decisionmaking and management. • Integration of socioeconomic, demographic, risk assessment, vulnerable populations and other important information into recovery planning and decisionmaking activities. • Increased community-wide support and understanding of sustainability and resiliency principles applicable to the opportunities presented during disaster recovery.

Function

The core recovery capability for community planning is the ability to effectively plan and implement disaster recovery activities, engaging the whole community to achieve their objectives and increase resilience. The Community Planning and Capacity Building RSF unifies and coordinates expertise and assistance programs from across the Federal Government to aid in restoring and improving the ability of Tribes, States and local governments to organize, plan, manage and implement recovery. The RSF assists States in developing a pre- and post-disaster system of support for their communities. This RSF also has an emphasis on integration of hazard mitigation throughout the continuum of pre- and post-disaster recovery planning and implementation. The RSF also serves as a forum for helping to integrate the nongovernmental and private sector resources into public sector recovery planning processes (Tables 2, 3 and 4).

RSF: ECONOMIC

Coordinating Agency: DOC
Primary Agencies: DHS/FEMA, DOC, DOL, SBA, TREAS, USDA
Supporting Organizations: CNCS, DOI, EPA, HHS

Mission

The mission of the Economic RSF is to integrate the expertise of the Federal Government to help local, State and Tribal governments and the private sector sustain and/or rebuild businesses and employment, and develop economic opportunities that result in sustainable and economically resilient communities after large-scale and catastrophic incidents.

Function

The core recovery capability for economic recovery is the ability to return economic and business activities (including agricultural) to a state of health and develop new economic opportunities that result in a sustainable and economically viable community. Economic recovery is a critical and integral part of recovery. Disasters not only damage property, but also entire markets for goods and services. The speed and effectiveness of returning a community to self-sufficiency and vitality depend upon quickly adapting to changed market conditions, reopening businesses and/or establishing new businesses. Businesses employ workers, provide for community needs and services and generate revenue once again, allowing the community, both its members and government, to provide for itself.

Considerable Federal funds are contributed to local, State and Tribal economic recovery as well as to other areas of recovery that necessarily strengthen the economy. The attraction of outside investment and the role of the private sector cannot be understated as foundational in a community's economic recovery. Thus, the role of the Economic RSF is to facilitate and enable that role by leveraging Federal resources, information and leadership. Informed management must accompany this capital investment to ensure its most effective use and compliance with all applicable Federal laws and regulations. This involves the coordination of Federal recovery programs and their integration with private sector efforts including those of nongovernmental and private volunteer organizations, nonprofits, investment capital firms and the banking industry.

The Economic RSF facilitates the progression from direct Federal financial assistance to community self-sustainment. Importantly, the RSF works closely with local community leadership who direct long-term economic recovery efforts. This requires the sustained engagement of possibly months or years by RSF leadership with the leadership of disaster-impacted jurisdictions. A complex undertaking, this RSF engages many entities utilizing government assistance as seed money. These actions encourage reinvestment and facilitate private-sector lending and borrowing necessary for the functioning of vital markets and economies. Effective economic recovery following a disaster is positively influenced by pre-disaster community planning including mitigation actions that increase community resilience. When coupled with informed decisions by local officials, it provides the confidence building necessary for economic recovery (Tables 5, 6 and 7).

RSF: Health and Social Services

Coordinating Agency: HHS.

Primary Agencies: CNCS, DHS (FEMA, NPPD & CRCL), DOI, DOJ, DOL, ED, EPA, VA

Supporting Organizations: DOT, SBA, TREAS, USDA, VA, ARC, NVOAD.

Mission

The Health and Social Services RSF mission is for the Federal Government to assist locally-led recovery efforts in the restoration of the public health, health care and social services networks to promote the resilience, health and well-being of affected individuals and communities.

Table 5. Pre-Disaster, the Economic RSF

- Identifies statutory, regulatory and policy issues that contribute to gaps, inconsistencies and unmet needs in economic recovery.
- Seeks innovative solutions to address preparedness, mitigation and resilience issues before a disaster strikes including comprehensive land use policy.
- Appreciates the value of community and economic development planning in disaster recovery; encourages and facilitates this planning through appropriate State government agencies.
- Develops initiatives and incentives to facilitate the integration of Federal efforts and resources with private capital and the business sector.
- Creates, encourages and participates in local, State, Tribal and Federal disaster recovery exercises to enhance skills and develop needed techniques.
- Leverages mitigation programs to create strong communities resilient to disaster.
- Works with local, State and Tribal officials to implement disaster resistant building codes and incentivize business and individual pre-disaster mitigation and preparedness activities.
- Seeks to promulgate resiliency policies and practices in agency programs and stakeholder operations, wherever appropriate.
- Sustains pre-disaster engagement activities possibly for months or years with the leadership of jurisdictions that may be impacted by a disaster.
- Encourages the establishment of disaster information networks for businesses.

Table 6. Post-Disaster, the Economic RSF

- When activated by the FDRC, the primary and supporting departments and agencies deploy in support of the Economic RSF mission.
- Works to apply and integrate plans developed pre-disaster to most effectively leverage Federal resources and available programs to meet local community recovery needs while aggressively integrating with the private sector to facilitate early and productive engagement.
- Develops an interagency action plan for each disaster to ensure the coordinated action of all Federal agencies, stakeholders and supporting entities in the support of local, State and Tribal governments.
- Incorporates mitigation measures into redevelopment following a disaster to build the community back stronger to minimize future risk.
- Building upon the relationships developed during pre-disaster planning, works closely with local community leadership during disaster recovery to provide technical assistance and data related to economic development.
- Maintains robust and accessible communications throughout the recovery process between the Federal Government and all other partners to ensure ongoing dialogue and information sharing.
- Engages the workforce development system, including State vocational rehabilitation programs, as a means of helping individuals who acquire a disability as part of the disaster return to work with the appropriate supports, accommodation and retraining (if necessary).

Table 7. Outcomes for the Economic RSF

Through the coordination of local, State, Tribal and Federal government programs and the private sector, the Economic RSF and local leadership leverages, following a disaster, community development plans and stakeholder relationships to create a new post-disaster economic condition meeting community needs. The following actions encourage reinvestment and facilitate private sector lending and borrowing necessary for the functioning of vital markets and economies. Sustained pre- and post-disaster mitigation actions create a community less at risk, strengthen future economic stability and create possible insurance benefits. Specific outcomes may include: • Workforce development initiatives are in place; jobs are created and retained. • Entrepreneurial and business development initiatives are in place. • Community-wide economic development plans are developed with broad input and consider regional economic recovery and resiliency. • Strategies for quickly adapting to changed market conditions, reopening businesses and/or establishing new businesses are in place. • Business initiatives to employ workers and generate revenue are in place. • Management plans ensure that the most effective use of Federal funds is in place. • Federal funds are withheld when discrimination on the basis of race, color, national origin, religion, sex, age, or disability are present. • Private and public sector actors have information they need to make informed decisions about recovery.

Function

The core recovery capability for health and social services is the ability to restore and improve health and social services networks to promote the resilience, health, independence and well being of the whole community. The Health and Social Services RSF outlines the Federal framework to support locally-led recovery efforts to address public health, health care facilities and coalitions, and essential social services needs. For the purposes of this RSF, the use of the term health will refer to and include public health, behavioral health and medical services. This Annex establishes (1) a Federal focal point for coordinating Federal recovery efforts specifically for health and social services needs; and, (2) a Federal operational framework outlining how Federal agencies plan to support local health and social services recovery efforts. This framework is flexible and can adjust during a disaster to complement local efforts, as needed (Tables 8, 9 and 10).

Table 8. Pre-Disaster, the Health and Social Services RSF

• Incorporates planning for the transition from response to recovery into preparedness and operational plans, in close collaboration with ESFs #3, #6, #8 and #11. • Incorporates planning for the transition from post-incident recovery operations back to a steady-state into preparedness and operational plans. • Develops strategies to address recovery issues for health, behavioral health and social services – particularly the needs of response and recovery workers, children, seniors, people living with disabilities, people with functional needs, people from diverse cultural origins, people with limited English proficiency and underserved populations. • Promotes the principles of sustainability, resilience and mitigation into preparedness and operational plans.

Table 9. Post-Disaster, the Health and Social Services RSF

- Maintains situational awareness to identify and mitigate potential recovery obstacles during the response phase.
- Leverages response, emergency protection measures4 and hazard mitigation resources during the response phase to expedite recovery5.
- Provides technical assistance in the form of impact analyses and supports recovery planning of public health, health care and human services infrastructure.
- Conducts Federal Health and Social Services RSF assessments with primary agencies.
- Identifies and coordinates Federal Health and Social Services RSF-specific missions with primary agencies6.
- When activated by the FDRC, the primary and supporting departments and agencies deploy in support of the Health and Social Services RSF mission, as appropriate.
- Establishes communication and information-sharing forum(s) for Health and Social Services RSF stakeholders with the State and/or community.
- Coordinates and leverages applicable Federal resources for health and social services.
- Develops and implements a plan to transition from Federal Health and Social Services recovery operations back to a steady-state.
- Identifies and coordinates with other local, State, Tribal and Federal partners to assess food, animal, water and air conditions to ensure safety.
- Evaluates the effectiveness of Federal Health and Social Services recovery efforts.
- Provides technical assistance in the form of impact analyses and recovery planning support of public health, health care, and human services infrastructure.
- Identifies and coordinates with other local, State, Tribal and Federal partners the assessment of food, animal, water and air conditions to ensure their safety.

Table 10. Outcomes for the Health and Social Services RSF

- Restore the capacity and resilience of essential health and social services to meet ongoing and emerging post-disaster community needs.
- Encourage behavioral health systems to meet the behavioral health needs of affected individuals, response and recovery workers, and the community.
- Promote self-sufficiency and continuity of the health and well-being of affected individuals; particularly the needs of children, seniors, people living with disabilities whose members may have additional functional needs, people from diverse origins, people with limited English proficiency, and underserved populations.
- Assist in the continuity of essential health and social services, including schools.
- Reconnect displaced populations with essential health and social services.
- Protect the health of the population and response and recovery workers from the longer- term effects of a post-disaster environment.
- Promote clear communications and public health messaging to provide accurate, appropriate and accessible information; ensure information is developed and disseminated in multiple mediums, multi-lingual formats, alternative formats, is age-appropriate and user-friendly and is accessible to underserved populations.

RSF: Housing

Coordinating Agency: HUD

Primary Agencies: DHS/FEMA, DOJ, HUD, USDA

Supporting Organizations: CNCS, DOC, DOE, EPA, HHS, SBA, U.S. Access Board, VA, ARC, NVOAD

Mission

Address pre- and post-disaster housing issues and coordinate and facilitate the delivery of Federal resources and activities to assist local, State and Tribal governments in the rehabilitation and reconstruction of destroyed and damaged housing, whenever feasible, and development of other new accessible, permanent housing options.

Function

The core recovery capability for housing is the ability to implement housing solutions that effectively support the needs of the whole community and contribute to its sustainability and resilience. Like infrastructure and safety services, housing is a critical and often challenging component of disaster recovery. It is critical because local economies cannot recover from devastating disasters without adequate housing, especially affordable housing. It is challenging because many years' worth of housing repair, rehabilitation, reconstruction and new construction often need to occur at an accelerated pace as a result of a disaster. These conditions create design, construction, labor, materials, logistics, inspection and financing issues.

The Housing RSF, through its member departments and agencies, works toward addressing disaster housing issues pre-disaster, focusing on solutions that are implementable, sustainable and resilient. As States and communities look to the Federal Government for assistance in housing both disaster survivors and others who choose to live in recovering communities, the Housing RSF coordinates and effectively integrates available housing-related resources, addresses conflicting policy and program issues and identifies gaps in service and assistance delivery.

Table 11. Pre-Disaster, the Housing RSF

• Works with local, State and Tribal governments, organizations and others in coordination with the *National Disaster Housing Task Force*, Joint Housing Solutions Group. • Identifies strategies and options that address a broad range of disaster housing issues such as those dealing with planning, zoning, design, production, logistics, codes and financing. • Builds accessibility, resilience, sustainability and mitigation measures into identified housing recovery strategies.

Table 12. Post-Disaster, the Housing RSF

• When activated by the FDRC, the primary and supporting departments and agencies deploy in support of the Housing RSF mission. • Coordinates and leverages Federal housing-related resources to assist local, State and Tribal governments to address housing-related, disaster recovery needs. • Encourages rapid and appropriate decisions regarding land use and housing location in the community or region. • Identifies gaps and coordinates a resolution of conflicting policy and program issues. • Maintains robust and accessible communications throughout the recovery process between the Federal Government and all other partners to ensure ongoing dialogue and information sharing.

Consistent with the *National Disaster Housing Strategy (NDHS)*, the Department of Homeland Security (DHS)/Federal Emergency Management Agency (FEMA) maintains lead responsibility for sheltering and interim housing with interim housing support from Housing and Urban Development (HUD) as well as other primary agencies and support organizations. Sheltering falls under ESF #6 in the *National Response Framework (NRF)* where DHS/FEMA is the coordinating agency. Interim housing, as its name implies, is a transition to permanent housing and is dependent on the period of transition as responsibility moves from Emergency Support Function (ESF) #6 to the Housing RSF. Addressing permanent housing, the third focus area of the NDHS, is under the Housing RSF (Tables 11, 12 and 13).

Table 13. Outcomes for the Housing RSF

Departments and agencies with expertise in long-term housing solutions work through this RSF and in conjunction with the *National Disaster Housing Task Force* so that:

- Housing resources that address local, State and Tribal disaster recovery housing needs are coordinated.
- Planning for current and post-disaster requirements are integrated into the organizations at the local and State level that perform land and community planning and building code administration.
- Local, State, Tribal and Federal programs, industry and construction options for addressing post-disaster housing needs are in place.
- Research results related to the disaster recovery housing area are shared.
- Interagency knowledge and expertise are shared with State-led housing task forces to address disaster housing issues.
- Pre- and post-disaster interaction and problem solving among Federal agencies and stakeholders with a focus on reconstructing permanent housing, including affordable and accessible housing that incorporates resilience, sustainability and mitigation concepts are facilitated.
- Timely construction of housing that complies with local, State and national model building codes, including accessibility standards, is facilitated.
- Loss of historic buildings and resources is minimized.

RSF: Infrastructure Systems

Coordinating Agency: DOD/USACE.
Primary Agencies: DHS (FEMA & NPPD), DOD/USACE , DOE , DOT.
Supporting Organizations: DHS, DOC, DOD, DOI, ED, EPA, FCC, GSA, HHS, NRC, TREAS, USDA, TVA.

Mission

Facilitate the integration of the capabilities of the Federal Government to support local, State and Tribal governments and other infrastructure owners and operators in their efforts to achieve recovery goals relating to the public engineering of the Nation's infrastructure systems.

Function

The core recovery capability for infrastructure systems is the ability to efficiently restore the infrastructure systems and services to support a viable, sustainable community and improves resilience to and protection from future hazards. The Infrastructure Systems RSF promotes a holistic approach to disaster recovery coordination, support, planning and implementation for infrastructure systems that serve the community. This includes single and multijurisdictional areas and regions.

The Infrastructure Systems RSF Coordinating Agency conducts operations in accordance with its authorities and resources to provide vital public engineering services to strengthen our Nation's security and reduce risks from disasters. When appropriate, the Coordinating Agency, working together with FEMA, facilitates and promotes the efforts of the RSF primary and supporting agencies to ensure those agencies with the requisite authorities, expertise, and resources are positioned to provide assistance to and collaborate with public and private sector infrastructure partners to the extent authorized by law. The Infrastructure Systems RSF Coordinating Agency does not directly undertake, however, any operational recovery or engineering activities outside the scope of its authorities and resources.

The Infrastructure Systems RSF serves as a collaborative forum for Federal Government engagement with local, State, Tribal and private sector representatives to focus on public engineering services that can reduce risks from disasters and expedite recovery. The collaborative efforts of this RSF involve government and private sector partners with expertise in public engineering services, as appropriate, across the infrastructure sectors identified through the *National Infrastructure Protection Plan (N I PP) Partnership Framework*[7]. Therefore, the scope of this RSF includes, but is not limited to, the following infrastructure sectors and subsectors: energy, water, dams, communications, transportation systems, Agriculture (food production and delivery), government facilities, utilities, sanitation, engineering, flood control and other systems that directly support the physical infrastructure of communities; as well as physical facilities that support essential services, such as public safety, emergency services and public recreation (Tables 14, 15 and 16).

RSF: Natural and Cultural Resources

Coordinating Agency: DOI.
Primary Agency: DHS/FEMA, DOI, EPA.
Supporting Organizations: ACHP, CNCS, CEQ, DOC, IMLS, LOC, NEA, NEH, USACE, USDA, Heritage Preservation

Mission

Integrate Federal assets and capabilities to help State and Tribal governments and communities address long-term environmental and cultural resource recovery needs after large-scale and catastrophic incidents.

Function

The core recovery capability for natural and cultural resources is the ability to protect natural and cultural resources and historic properties through appropriate response and recovery actions to preserve, conserve, rehabilitate, and restore them consistent with post-disaster community priorities and in compliance with appropriate environmental and cultural resources laws. The Natural and Cultural Resources RSF coordinates departments and agencies working together to provide information and assistance to communities seeking to preserve, protect, conserve, rehabilitate, recover and restore natural and cultural resources during recovery.

Table 14. Pre-Disaster, the Infrastructure Systems RSF

- Develops guidance and standard procedures for rapid activation of RSF capabilities to support community recovery.
- Identifies relevant statutory and/or regulatory programs, potential capabilities and/or limiting factors pertaining to recovery support for infrastructure systems.
- Provides a forum for interagency coordination, information sharing and exchange of effective practices.
- Supports planning, preparedness, education, training and outreach efforts to enhance capabilities for recovery.
- Works with partners to identify critical facilities and ensure considerations are made to reduce risk pre- and post-disaster.

Table 15. Post-Disaster, the Infrastructure Systems RSF

- When activated by the FDRC, the primary and supporting departments and agencies deploy in support of the Infrastructure Systems RSF mission.
- Supports the recovery of infrastructure systems, dependent on the nature and scope of the disaster, and the specific authorities and programs within the jurisdiction of participating departments and agencies.
- Participates in the national-level coordination of damage and community needs assessments as appropriate to ensure infrastructure considerations integrate into the post- disaster public and private sector community planning process.
- Deploys RSF resources, as required by the specific disaster situation and consistent with the specific authorities and programs of the participating departments and agencies, to the field to assist the affected community in developing an Infrastructure Systems Recovery action plan that:
  - Avoids the redundant, counterproductive, or unauthorized use of limited capital resources necessary for infrastructure/recovery.
  - Helps resolve conflicts, including those across jurisdictional lines, resulting from the competition for key resources essential to infrastructure systems recovery[8].
  - Sets a firm schedule and sequenced time structure for future infrastructure recovery projects.
- Works with RSF partners to leverage available financial and technical assistance, both from governmental and nongovernmental sources, in the execution of the community's Infrastructure Systems Recovery action plan.
- Promotes rebuilding infrastructure in a manner which will reduce vulnerability to future disasters impacts.
- Maintains robust and accessible communications throughout the recovery process between the Federal Government and all other partners to ensure ongoing dialogue and information sharing.

Table 16. Outcomes for the Infrastructure Systems RSF

The Infrastructure Systems RSF provides the coordinating structures, framework and guidance to ensure:

- Resilience, sustainability and mitigation are incorporated as part of the design for infrastructure systems and as part of the community's capital planning process.
- Infrastructure systems are fully recovered in a timely and efficient manner to minimize the impact of service disruptions. The private sector critical infrastructure has the incentive and the means to support a unified community and national recovery effort.
- The capacity of all infrastructure systems is adequately matched to the community's current and projected demand on its built and virtual environment.

Relevant agencies and partners are those with expertise and programs including, but not limited to, specific natural and cultural resource issue identification, assessment and management (e.g., fish and wildlife, historic and traditional cultural properties, hydrology); natural and cultural resource planning; environmental planning and historic preservation compliance under Federal laws and *Executive Orders* (specific to programs that provide funding for disaster recovery); and community sustainability (Tables 17, 18 and 19).

Pre-Disaster, the Natural and Cultural Resources RSF

- Identifies relevant Federal programs and incentives that have a role in supporting the preservation, protection, conservation, rehabilitation, recovery and restoration of natural and cultural resources during recovery.
- Develops a pre-disaster Natural and Cultural Resources RSF action plan to identify and communicate priority actions.
- Identifies and prioritizes gaps and inconsistencies within and between relevant Federal regulations, policies, program requirements and processes affecting natural and cultural resources that are used in disaster recovery, either separately or in combination with one another, and makes recommendations to the National Disaster Recovery Planning (NDRP) Division at FEMA Headquarters and specific Federal agencies.
- Works with private nonprofits and other nongovernmental organizations (NGOs) to leverage opportunities to encourage local, State and Tribal governments and institutions to develop emergency management plans that integrate natural and cultural resource issues.
- Promotes the principles of sustainable and disaster resistant communities through the protection of natural resources such as coastal barriers and zones, floodplains, wetlands and other natural resources critical to risk reduction.
- Assesses appropriate hazard mitigation strategies for the protection of cultural resources.

9. Planning for Successful Disaster Recovery

Proper pre- and post-disaster planning is a prerequisite for the implementation of a well-orchestrated recovery process at the local, State and Tribal levels. Preparedness initiatives help guide the recovery process to effectively and efficiently achieve a community's disaster recovery priorities. Both pre- and post-disaster recovery planning are critical for communities to develop resilience and for successful and timely recovery.

Table 18. Post-Disaster, the Natural and Cultural Resources RSF

• When activated by the FDRC, the primary and supporting departments and agencies deploy in support of the Natural and Cultural Resources RSF mission. • Works to leverage Federal resources and available programs to meet local community recovery needs. • Identifies opportunities to leverage natural and cultural resource protection with hazard mitigation strategies. • Addresses government policy and agency program issues, gaps and inconsistencies related to natural and cultural resource issues. • Coordinates cross-jurisdictional or multistate and/or regional natural and cultural resource issues to ensure consistency of Federal support where needed. • Encourages responsible agencies at all levels of government and their important private sector partners to support the local community's recovery plan and priorities by developing a Natural and Cultural Resources action plan that identifies how the agencies leverage resources and capabilities to meet the community's needs. • Synchronizes the Natural and Cultural Resources action plan with other RSFs, as appropriate to support the broader vision of Federal support to disaster recovery. • Helps communities and State and Tribal governments to leverage opportunities inherent in recovery to mitigate impacts to environmental or cultural resources. • Promotes a systematic, interdisciplinary approach to understand the interdependencies and complex relationships of the natural and cultural environments. • Maintains robust and accessible communications throughout the recovery process between the Federal Government and all other partners to ensure ongoing dialogue and information sharing.

Table 19. Outcomes for the Natural and Cultural Resources RSF

With expertise drawn from Federal departments and agencies, the Natural and Cultural Resources RSF works so that: • Considerations related to the management and protection of natural and cultural resources and historic properties (NCH) resources, community sustainability and compliance with environmental planning and historic preservation requirements are integrated into recovery. • Local communities, States and Tribal governments are ready to address post-disaster natural and cultural resource recovery needs. • Programs to support disaster recovery, coordination of technical assistance and capabilities and data sharing are coordinated. • Natural and cultural assessments and studies needed post-disaster, including proposed solutions to environmental and historic preservation policy and process impediments, are developed.

Pre-Disaster Recovery Planning

Pre-disaster recovery planning enables local, State and Tribal governments to effectively direct recovery activities and expedite a unified recovery effort. Pre- disaster plans provide a common platform to guide recovery decisions and activities. When done in conjunction with local and regional comprehensive and community development, pre-disaster planning helps to identify recovery priorities, incorporate hazard mitigation strategies in the wake of a disaster and articulate post-disaster options. By integrating and coordinating planning initiatives, a community further increases local resilience.

In addition to the general elements of the pre-disaster planning process (Table 20), there are also elements specific to the various participants in the process. The responsibility of preparing for disaster recovery begins with the individual and builds to the larger responsibility of the community and local government. Community planning efforts are supported by voluntary, faith-based and community organizations; businesses; and local, State, Tribal and Federal governments. Details on pre- and post-disaster planning activities can be found in Appendix C.

Pre-Disaster Recovery Planning Elements

Pre-disaster recovery planning involves a State or community articulating a process for how it organizes and manages its recovery, establishes relationships among stakeholders and develops methods for prioritizing recovery decisions and land use considerations. Elements of a pre-disaster recovery planning and coordination system (Table 21) may include the following:

Table 20. Key Principles of Pre-Disaster Recovery Planning

Pre-disaster planning relies on key principles to:

- Establish clear leadership, coordination and decisionmaking structures at the local, State and Tribal levels.
- Develop pre-disaster partnerships to ensure engagement of all potential resources through the following methods:
  - Identify and engage stakeholders including, but not limited to, the general public, community leaders, faith-based organizations, nonprofit organizations and private sector entities.
  - Organize connections to interface with local, State, Tribal and Federal governments.
  - Ensure community participation of historically underserved populations including diverse racial and ethnic communities, individuals with disabilities and others with access and functional needs, children, seniors, and individuals with limited English proficiency.
- Test and evaluate pre-disaster plans through seminars, workshops and exercises.
- Build partnerships between neighborhoods and local government agencies that form the basis for pre-and post-multihazard assessments and support for mitigation actions.
- Integrate pre-disaster recovery planning (e.g., response, land use and hazard mitigation planning) with other appropriate community planning (e.g., comprehensive, accessibility design and capital improvement planning).
- Identify limitations in community recovery capacity and the means to supplement this capacity.
- Incorporate sustainable development, including environmental, historic preservation and financial elements, into recovery planning guidelines.
- Develop an accessible public information campaign that addresses the concerns of the public and an array of possible scenarios.
- Prepare pre-disaster *Memoranda of Understanding* (M OUs) as a way to establish early partnerships, planning initiatives and expectations with stakeholders, community faith- based organizations, nonprofit groups and private sector entities.
- Develop and implement recovery training and education as a tool for building recovery capacity and making it available to all other stakeholders.
- Identify resource requirements and conduct acquisition planning.

Table 21. PRE-Disaster Recovery Planning Recommended Activities

Assessment
• Identify hazards, assess risks and vulnerabilities. • Identify limitations in recovery capacity, and means to supplement this capacity. • Identify areas of potential financial challenges.
Communication and Outreach.
• Identify strategies to use in the development of the pre-disaster recovery planning process. • Develop outreach and communications strategies for use during post-disaster recovery. • Ensure community participation of underserved and disadvantaged populations including the use of alternative communications formats and multiple languages. • Ensure effective communications for all participants, including individuals with disabilities and individuals with limited English proficiency.
Stakeholders
• Identify sectors of the community to participate in pre- and post-disaster recovery planning and coordination.
Partnerships
• Develop pre-disaster partnerships that ensure engagement of all potential resources and issues. • Encourage full engagement of the public and recovery stakeholders. • Organize connections and interface with the local government.
Guiding Principles and Recovery Priorities
• Determine principles to guide recovery decisionmaking. • Explore how priorities are determined following a disaster. • Incorporate sustainability into overall planning guidance.
Organizational Framework
• Establish clear leadership, coordination and decisionmaking structures throughout all levels of government.
Concept of Operations
• Establish the operational framework that is followed immediately after a disaster occurs. • Establish maintenance procedures for updating pre- and post-disaster recovery plans.
Process for Post-Disaster Recovery Planning

• Clearly articulate the connectivity between mitigation, comprehensive and regional
• sustainability planning and other policy positions.
• Identify how the community will work together after a disaster to develop their plan for recovery
• Use a multihazard approach to recovery planning and preparedness.
• Identify priority recovery and redevelopment activities.
• Organize decisions through the use of a planning system that:
  • Evaluates the likely conditions and needs after a disaster.
  • Sets recovery goals and objectives.
  • Measures progress against those goals and objectives.

Exercise
• Test pre-disaster planning, preparation and staff capabilities by implementing recovery exercises. • Evaluate performance and revise pre-disaster recovery plans accordingly.

Planning Considerations
• Identify specific planning considerations that must be taken into account in the development of a recovery plan, including but not limited to, place-based mitigation issues such as: • Wild/rural/urban interfaces. • Floodplain management. • Coastal zones. • Seismic areas. • Historic and cultural properties, districts, landscapes, and traditional cultural properties.

Post-Disaster Recovery Planning

Communities impacted by a disaster should develop a process for optimally managing their recovery effort and resources. Post- disaster community recovery planning (Table 22) puts complex decisions in the context of the disaster and forms the foundation for allocating resources. The planning process provides the benchmark to measure the affected community's progress towards a successful outcome.

Table 22. Key Principles of Post-Disaster Recovery Planning

All disaster-impacted communities can benefit by engaging in disaster recovery planning and creating plans that are meaningful to multiple audiences, including potential funders, Tribal governments, State and Federal level agencies and members of the community. The post-disaster planning process: • Organizes recovery priorities and tasks through the use of a planning process to: • Evaluate the conditions and needs after a disaster. • Assess risk. • Set goals and objectives. • Identify opportunities to build in future resilience through mitigation. • Identify specific projects in areas of critical importance to the community's overall recovery. • Uses a community-driven and locally managed process, designed to promote local decisionmaking and ownership of the recovery planning and implementation effort. • Works collaboratively with all groups of people affected by the disaster to promote inclusive and accessible outreach to their communities and address issues relevant to them. Ensures inclusion and encourages participation of individuals and communities that may require alternative and/or additional outreach support (e.g., racial/ethnic communities, individuals with limited English proficiency, and people with disabilities). • Incorporates considerations that include the concept of "growing smarter" as the recovery continuum progresses. This includes compliance with standards for sustainable and accessible design, alteration and construction. • Integrates multihazard considerations into mitigation and preparedness activities. • Builds partnerships among local agencies, jurisdictions and State, Tribal and Federal governments. • Provides well-defined activities and outcomes — including schedules and milestones — aimed at achieving recovery. • Develops tools and metrics for evaluating progress against set goals, objectives and milestones. • Identifies resource requirements and conducts acquisition planning.

Planning for the complex needs of the full community and bringing all stakeholders to a common planning table with a commitment to physical, programmatic and communications accessibility helps create a successful post-disaster recovery process. A significant challenge of post-disaster recovery planning is developing a plan quickly enough to meet the needs of residents and businesses. The post-disaster planning process operates on a much faster timeline than traditional or pre-disaster planning processes. However, one of the basic goals of the process is to develop the relationships and interagency cooperation that continue to serve the recovery process once planning is complete.

Suggested Post-Disaster Recovery Planning Elements

A post-disaster plan is a discrete process that produces a document or series of documents for the disaster at hand that results in integrated recovery and reconstruction programs, actions and recovery. This post-disaster recovery plan is based on the strategy and process laid out in the pre-disaster plan. Post-disaster plans guide funding for a wide variety of public sector and nongovernmental investments.

The post-disaster recovery planning process brings the community together to develop an overall framework for coordination and recovery planning (Table 23). The process assists States and communities in focusing on recovery issues and needs, developing projects and strategies to address those needs and determining measures of success to better manage recovery. Organizing and managing the recovery process allows a community to take advantage of the opportunities created by recovery. Each community determines its process for post- disaster recovery planning. General elements that may be helpful when considering the development of a post-disaster recovery plan include the following:

Table 23. Post-Disaster Recovery Planning Elements[9]

Assessment.
• Assesstheneedcreatedbythedisastertodeterminewhererecoveryissuesarepresentgeographically by sector (e.g., housing, health care, infrastructure, environment, economy). • Determine areas of future risk and mitigation opportunities, such as reviewing past Federally-funded mitigation activities for effectiveness and use as baseline for new recovery planning and assessment needs.
Leadership.
• Identifyanindividualorgroupaswellassupportingstructuresrequiredtoleadtheprocess in a manner that complies with all relevant laws, including civil rights laws.
Support
• Coordinate with all community leaders to ensure participation and validity of the process. • Identify outside resources, financial and technical, that provide support to the overall recovery effort.
Communication and Outreach
• Establish an accessible process for exchanging information between the public and leadership. • Develop a communications map to ensure all sectors of the community are engaged in the process. • Use nontraditional communications outlets to reach as much of the community as possible. • Ensure effective communications for all participants, including individuals with disabilities and individuals with limited English proficiency.

Existing Guidance Documents
• Use mitigation, comprehensive and other community plans to guide the identification of priority redevelopment and reconstruction within recovery.
Build Consensus
• Work together to move recovery forward. • Continue to engage the community and reach out to new stakeholders. • Identify and address conflict.
Recovery Issues
• Determine the areas of concern and the impact these areas have on recovery. • Identify areas of opportunity in recovery planning.
Recovery Vision and Goals
• Identify areas that strengthen and revitalize the community. • Develop and document the recovery vision and goals. • Solicit public participation in the development and confirmation of the vision and goals.
Strategy
• Use existing planning documents on hazard mitigation, comprehensive and regional planning to develop strategies. • Develop projects and programs to meet the recovery vision and goals created by the community. • Evaluate projects and programs to determine their impact on recovery, feasibility, public support, sustainability initiatives, effective use of resources and other criteria as determined by the community.
Plan Writing
• Document the vision, goals, projects and programs. • Provide a draft to stakeholders. • Revise based on feedback.
Implementation
• Determine the implementation plan and priorities for recovery projects. • Identify key leaders responsible for undertaking next steps or moving recovery strategies forward. • Remain flexible.
Funding
• Use existing public and private resources and new funding streams to creatively package resources. • Phase large projects to allow for more flexible application of funding.
Update
• Revise plans as needed to meet changing recovery needs and priorities. • Continue to update and engage the broader stakeholder population on next steps. • Monitor progress and convey achievements to all stakeholders.

10. Community Considerations

Communities across the country vary in size, geography and demographics. A successful recovery effort takes these various community considerations into account. What follows is guidance on accessibility and recovery, ensuring inclusion of all community members, unmet needs, rural recovery needs, and high density urban area needs.

Accessibility and Recovery

The guidance included here is specific to issues related to children and adults with physical, mental, cognitive, intellectual and sensory disabilities as well as others with access and functional needs.

Recognize that best practices are not to be confused with legal obligations to engage in recovery activities that are fully inclusive of individuals with disabilities and other individuals with access and functional needs. Those applying the NDRF should be aware of statutory and executive order obligations involved, which may include:

- *Robert T. Stafford Disaster Relief and Emergency Assistance Act (Stafford Act)*, as amended.
- *Post-Katrina Emergency Management Reform Act.*
- *Rehabilitation Act of 1973,* as amended.
- *Americans with Disabilities Act (ADA)*, as amended 2008.
- *Fair Housing Act of 1968*, as amended.
- *Architectural Barriers Act of 1968.*
- *Communications Act of 1934*, as amended.
- *Individuals with Disabilities Education Act (IDEA) of 1975*, as amended.
- Title V I of the *Civil Rights Act of 1964*, as amended.
- The *Age Discrimination Act of 1975.*
- Executive Order 13347, *Individuals with Disabilities in Emergency Preparedness.*

These statutory and executive order obligations include accessibility in architecture, transportation, housing and effective communications, employment, education, policies and programs including those receiving Federal funding. Strategies for fulfilling these obligations include:

- Ensure the integration of people with disabilities and other individuals with access and functional needs into all aspects of emergency management rather than as a supplement or special plan in, among others, policies, practices, procedures, guidelines, standards, *Memoranda of Understanding* (M O Us) and agreements or contracts.
- Use existing resources to determine accessibility of facilities and programs. Determine shortfalls based on tools and address deficiencies.
- Engage in pre-disaster contracting and planning to meet the emergency needs of children and adults with disabilities, including the provision of disability- related assistance and functional needs support services, consumable medical supplies, durable medical equipment, accessible transportation and accessible housing.
- Involve community, cultural and disability organizations, such as independent living organizations, protection and advocacy agencies and disability agencies in recovery planning efforts and all recovery committee types. Integrate disability and access and functional needs considerations into housing, economic and workplace development, health care, child care, transportation and infrastructure strategies.

- Conduct disaster recovery awareness training for stakeholders, including disability navigators; advocacy organizations, including those representing the needs of children with disabilities, individuals from diverse cultural origins, child care providers and schools; senior centers and aging agencies; rehabilitation offices; medical authorities and relevant organizations. Provide training that is legally compliant by using tools such as the *ADA Best Practices Tool Kit.*
- Consider that recovery partners (including emergency managers, social workers, organizations and agencies) may lack the familiarity that is presumed necessary for working with children and adults with disabilities and others with access and functional needs and the organizations that support these populations.
- Recognize that there are individuals who have acquired or exacerbated disabilities as a result of the disaster. These individuals may need added assistance to familiarize themselves with the processes to access services and support so that they can be as independent as possible and participate in the recovery process.
- Consider the continuing impact of a hazard, such as lingering smoke or the long-term effects of debris, on children and adults with existing and new disabilities.
- Ensure that all print, electronic and face- to-face communications are accessible to people with disabilities and other at-risk populations with access and functional needs. Provide necessary auxiliary aids and services to achieve effective communications, but not limited to, interpreters, computer-assisted real-time transcription, large print, captioning, audio descriptions, wayfinding and note taking.
- Ensure that material relevant to disabilities and emergency management is accessible and available along with other electronic material.
- Explore ways to use accessible and multi-lingual social media tools to collect and disseminate disaster and recovery information.
- Liaise with large-scale employers and providers of vocational and job training support to address the recovery of the employment sector.
- Ensure that recovery measures and metrics take into account the recovery progress of persons with disabilities and other individuals with access and functional needs.

Ensuring Inclusion of All Community Members

To be successful, recovery measures and metrics must take into account the needs of all community members. Actions, both intentional and unintentional, that exclude groups of people based on race, color, national origin, sex, age or disability, can have long-term negative consequences on entire communities and may violate the law. Understanding legal obligations and sharing best practices when planning and implementing recovery strategies to avoid excluding groups on these bases is critical.

As with the Accessibility and Recovery section, those who are engaging in recovery activities are covered by specific legal obligations that prohibit discrimination on the basis of race, color or national origin, including limited English proficiency, sex and age. These statutory and legal obligations include:

- Title VI of the *Civil Rights Act of 1964*, which prohibits discrimination or the denial of benefits on the basis of race, color or national origin, including limited English proficiency, under any program or activity receiving Federal financial assistance;
- *Robert T. Stafford Disaster Relief and Emergency Assistance Act (Stafford Act)*, which prohibits discrimination on the basis of race, color, religion, nationality, sex, age, disability, English proficiency or economic status;
- (1) Executive Order 13166, *Improving Access to Services for Persons with Limited English Proficiency*, which requires Federal agencies to take reasonable steps to ensure meaningful access to limited English proficient individuals in their Federally conducted activities. The Executive Order also directs Federal agencies to provide Title VI guidance to recipients of Federal funds regarding the requirement to take reasonable steps to provide meaningful access to limited English proficient individuals;
- *Fair Housing Act*, which prohibits discrimination in housing based on race, color, national origin, sex, religion, disability and familial status by local, State and Federal government and private housing providers, and which requires recipients of Federal housing funds to take affirmative steps to promote fair housing.
- *Age Discrimination Act of 1975*, which prohibits discrimination on the basis of age under any program or activity receiving Federal financial assistance.
- *Executive Order 12898 (February 11, 1994) Federal Actions to Address Environmental Justice in Minority Populations and Low-Income Populations.* Directs that each Federal agency shall make achieving environmental justice part of its mission by identifying and addressing, as appropriate, disproportionately high and adverse human health or environmental effects of its programs, policies and activities on minority populations and low-income populations.

Intentional race, color, national origin, disability, sex or age discrimination in the conduct of recovery efforts clearly violates civil rights laws. Recipients of Federal funds, including local and State governments and private housing providers, are also prohibited from implementing neutral policies and practices that have a discriminatory impact based on these characteristics. Finally both Federal agencies and recipients must take reasonable steps to provide meaningful access to limited English proficient individuals.

Promising practices to avoid discriminatory consequences in recovery efforts include:

- Conducting outreach to community organizations serving racially and ethnically diverse populations for purposes of advising them about available assistance, eligibility standards for assistance and including them in disaster preparedness, recovery and reconstruction planning.
- Ensuring that the location for meetings seeking public input is accessible to all impacted community members and advertisement for such meetings take into account the various ways in which different members access such information.
- Making sure that language services — including bilingual staff, interpreters and translated materials — are in place so that limited English proficient persons are able to communicate their needs, apply for assistance and receive important information about the process.

- Taking steps to encourage all eligible persons in need to come forward to receive vital disaster benefits; not all benefits require a person to be legally present and some allow children of undocumented parents to receive aid. Sharing information about the minimum eligibility standards for obtaining benefits, so that eligible individuals, including eligible immigrants and/or their children can access aid without difficulty or fear.
- Not applying more stringent eligibility criteria for recovery aid based on race, color, national origin, language or other prohibited bases.
- Ensuring affordable rental housing in racially and ethnically diverse communities and seeking input from those communities regarding rebuilding efforts.
- Monitoring and addressing possible housing discrimination against displaced persons seeking a place to live after a disaster. Complaint procedures should be clearly articulated and nondiscrimination policies should be in place.
- Ensuring that temporary homes are adequate and conveniently located to essential services such as schools.
- Planning for how residents may access alternative medical facilities, particularly via public transportation, if local facilities are damaged or destroyed. Engage impacted community members in the planning for rebuilding these facilities.
- Ensuring that recovery services are accessible by public transportation. Plan for community access to recovery services in the event public transportation systems are disrupted.
- Monitoring and addressing possible recovery-related fraud, particularly fraud targeted at racially and ethnically diverse communities.
- Prohibiting discriminatory terms or conditions or discrimination in housing services (for example, higher security deposits or higher rent for tenants based on their race, color or national origin).

Unmet Needs

A successful recovery plan adequately attends to and addresses unmet needs for individuals and families as they recover from a disaster. This is particularly true for vulnerable and underserved populations. As part of the post- disaster recovery needs assessment, recovery planners survey and interview community members to obtain first-hand information on unmet needs. Recovery planners subsequently measure those needs against the roles, responsibilities and capabilities of government and other stakeholders. Recovery planners collaborate closely with the Voluntary Agency Liaisons (VALs) and other applicable government offices. They use disaster research to facilitate consideration and inclusion of unmet needs throughout the recovery planning process. Typical areas of enduring need after a disaster include:

- Long-term mental and behavioral health concerns for children and adults in relation to traumatic events induced or exacerbated by the disaster.
- Transportation for and during relocation.

- Long-term housing including housing that recognizes the need for accessibility and affordability.
- Comprehensive case management.
- Children's stability within schools and child care settings.
- Investigation of underinsured properties.
- Affordability of home repairs or insurance deductibles.
- Legal services.
- Middle-class families who are outside the purview of traditional poverty relief or may be either ineligible or unaware of programs offered through agencies that assist the poor.
- Accessible financial assistance, including low interest loans for credit-challenged individuals.
- Access to information and communications technologies including internet, voice and broadcast technologies.

Rural Area Recovery Needs

Rural communities have particular needs following a disaster. Factors that may impact recovery in rural areas include: lack of local resources or management; large stretches of land that are thinly populated; and infrastructure dispersed across a vast stretch of land. These factors may make logistical endeavors challenging. Such communities may need a pre-disaster regional recovery plan but may lack the resources, leadership or political autonomy to engage in one.

In rural communities, sometimes the human population incurs the greatest loss through devastation to the land itself, rather than any man-made assets. Whether the disaster loss is caused through drought, floods, crop blight or livestock illnesses, agricultural concerns demand a recovery approach that is distinct from densely populated urban settlements or manufacturing centers. Other communities needing a rural recovery approach are those dependent upon such industries as forestry, mining, fisheries or oil/mineral exploration. In addition, Tribal governments may be rural in nature, suffering disproportionately in a disaster due to a similar dependence on land.

High-Density Urban Area Recovery Needs

Large-scale disasters sometimes fundamentally change the landscape of urban communities, including the social, business and physical landscapes. New urban plans need to be created post-disaster and community members need to be involved in the process. Community leaders and members alike assess the new challenges and opportunities that the community faces and create a preferred future for the urban area that may not be the same as simply returning to pre-disaster conditions.

11. ABBREVIATIONS

Departments, Organizations and Abbreviations

ADA	*Americans with Disabilities Act*, as amended 2008
ACHP	Advisory Council on Historic Preservation
ARC	American Red Cross
CEQ	Council on Environmental Quality
CIKR	Critical Infrastructure and Key Resources
CNCS	Corporation for National and Community Service
COG	Continuity of Governments
COOP	Continuity of Operations
CRCL	Office for Civil Rights and Civil Liberties
DA	Disaster Assistance
DHS	Department of Homeland Security
DOC	Department of Commerce
DOD	Department of Defense
DOE	Department of Energy
DOI	Department of the Interior
DOJ	Department of Justice
DOL	Department of Labor
DOT	Department of Transportation
DPA	Defense Production Act
ED	Department of Education
EEOC	Equal Employment Opportunity Commission
EPA	Environmental Protection Agency
ESF	Emergency Support Function
FCC	Federal Communications Commission
FCO	Federal Coordinating Officer
FDRC	Federal Disaster Recovery Coordinator
FEMA	Federal Emergency Management Agency
GSA	General Services Administration
HUD	Department of Housing and Urban Development
HHS	Department of Health and Human Services
IDEA	*Individuals with Disabilities Education Act*
IMLS	Institute of Museum and Library Services
JFO	Joint Field Office
LDRM	Local Disaster Recovery Manager
LOC	Library of Congress
LTCR	Long-Term Community Recovery
LTDRO	Long-Term Disaster Recovery Office
MOU	*Memorandum of Understanding*
NCD	National Council on Disability
NCH	natural and cultural resources and historic properties
NCP	National Continuity Plan

NDHS	*National Disaster Housing Strategy*
NDHTF	National Disaster Housing Task Force
NDRF	*National Disaster Recovery Framework*
NDRP	National Disaster Recovery Planning
NDRPD	National Disaster Recovery Program Database
NEA	National Endowment for the Arts
NEH	National Endowment for the Humanities
NFIP	National Flood Insurance Program
NGO	Nongovernmental Organization
NIMS	National Incident Management System
NIPP	*National Infrastructure Protection Plan*
NLRB	National Labor Relations Board
NPPD	National Protection Programs Directorate
NRC	Nuclear Regulatory Commission
NRF	*National Response Framework*
NVOAD	National Voluntary Organizations Active in Disaster
OFAs	other Federal agencies
PKEMRA	Post-Katrina Emergency Management Act
POC	point of contact
RISC	Regional Interagency Steering Committee
RSF	Recovery Support Function
SBA	Small Business Administration
SCO	State Coordinating Officer
SDRC	State Disaster Recovery Coordinator
SME	Subject Matter Expert
SOP	Standard Operating Procedure
TDRC	Tribal Disaster Recovery Coordinator
TREAS	Department of the Treasury
TVA	Tennessee Valley Authority
U.S. ACCESS BOARD	United States Access Board
USACE	U.S. Army Corps of Engineers
USDA	U.S. Department of Agriculture
USGS	United States Geological Survey
VA	Department of Veterans Affairs
VAL	Voluntary Agency Liaison
VALS	Voluntary Agency Liaison Specialist

12. DEFINITIONS

Access and Functional Needs – Persons who may have additional needs before, during and after an incident in functional areas, including but not limited to: maintaining independence, communication, transportation, supervision, and medical care. Individuals in need of additional response assistance may include those who have disabilities; live in institutionalized settings; are seniors; are children; are from diverse cultures; have limited English proficiency or are non-English speaking; or are transportation disadvantaged.

Access/Accessible – The suitability or adaptability of programs, services, activities, goods, facilities, privileges, advantages or accommodations provided by a public or private (for-profit or not-for-profit) entity, or by any entity to which it contracts for all members of the population, including individuals with disabilities.

Capacity – A combination of all the strengths and resources available within a community, society or organization that can reduce the level of risk, or the effects of a disaster. (From the *UN International Strategy for Disaster Reduction.*)

Capacity Building – Efforts aimed to develop human skills or societal infrastructure within a community or organization needed to reduce the level of risk, or the effects of a disaster. (From the *U.N. International Strategy for Disaster Reduction.)*

Catastrophic Incident – Any natural or man-made incident, including terrorism that results in extraordinary levels of mass casualties, damage, or disruption severely affecting the population, infrastructure, environment, economy, national morale, and/or government functions. A catastrophic event could result in sustained national impacts over a prolonged period of time; almost immediately exceeds resources normally available to local, State, Tribal, and private sector authorities in the impacted area; and significantly interrupts governmental operations and emergency services to such an extent that national security could be threatened.

Community – A network of individuals and families, businesses, governmental and nongovernmental organizations and other civic organizations that reside or operate within a shared geographical boundary and may be represented by a common political leadership at a regional, county, municipal or neighborhood level.

Consumable Medical Supplies – Generally nonprescribed, nondurable, disposable, single-use medical supplies that are most beneficial to persons with a disability, illness, injury or functional need to maintain their level of independence.

Critical Infrastructure – Systems and assets, whether physical or virtual, so vital that the incapacity or destruction of such may have a debilitating impact on the security, economy, public health or safety, environment, or any combination of these matters, across any local, State, Tribal and Federal jurisdiction.

Cultural Resources – Aspects of a cultural system that are valued by or significantly representative of a culture or that contain significant information about a culture. Cultural resources may be tangible entities or cultural practices. Tangible cultural resources are categorized as districts, sites, buildings, structures, and objects for the National Register of Historic Places and as archeological resources, cultural landscapes, structures, museum objects and archives, and ethnographic resources for Federal management purposes. Also includes cultural items as that term is defined in section 2(3) of the *Native American Graves Protection and Repatriation Act* [25 USC 3001(3)]; and archeological resources, as that term is defined in section 3(1) of the *Archaeological Resources Protection Act* of 1979 [16 USC 470bb(1)].

Debris - The remains of something broken down or destroyed.

Durable Medical Equipment – Multiuse medical equipment for the benefit of a person who has an illness, injury, disability or functional need to maintain their level of independence.

Functional Needs – The needs of an individual who under usual circumstances is able to function on their own or with support systems. However, during an emergency, their level of independence is challenged.

Historic Properties – Any prehistoric or historic district, site, building, structure, or object included in, or eligible for inclusion in the National Register of Historic Places, including artifacts, records and material remains which are related to such district, site, building, structure, or object. [(16 USC Section 70(w)(5)].

Individual with Disability – The term refers to a person (child or adult) who has a physical or mental impairment that substantially limits one or more major life activities; a person who has a history or record of such impairment; or a person who is perceived by others as having such impairment. The term "disability" has the same meaning as that used in the *Americans with Disabilities Act (ADA) Amendments Act of 2008, P.L. 110 – 325*, as incorporated into the ADA. See http://www.ada.gov/pubs/ada.htm for the definition and specific changes to the text of the ADA. State laws and local ordinances may also include individuals outside the Federal definition. Children and adults may have physical, sensory, mental health, cognitive and/or intellectual disabilities resulting in access and functional needs and may require assistance to maintain independence.

Individual with Limited English Proficiency – The term refers to an individual who does not speak English as his/her primary language and who has a limited ability to read, write, speak or understand English.

Intermediate Recovery – Phase of recovery which involves returning individuals, families, critical infrastructure and essential government or commercial services to a functional, if not pre-disaster, state. Such activities are often characterized by temporary actions that provide a bridge to permanent measures.

Long-Term Recovery – Phase of recovery that may continue for months or years and addresses complete redevelopment and revitalization of the impacted area, rebuilding or relocating damaged or destroyed social, economic, natural and built environments and a move to self-sufficiency, sustainability and resilience.

Major Disaster – As defined by the Stafford Act, any natural catastrophe (including any hurricane, tornado, storm, high water, wind-driven water, tidal wave, tsunami, earthquake, volcanic eruption, landslide, mudslide, snowstorm, or drought) or, regardless of cause, any fire, flood or explosion, in any part of the United States, which in the determination of the President causes damage of sufficient severity and magnitude to warrant major disaster assistance under this act to supplement the efforts and available resources of local, State governments and disaster relief organizations in alleviating the damage, loss, hardship or suffering caused thereby.

Mitigation – Capabilities necessary to reduce loss of life and property by lessening the impact of disasters. Mitigation capabilities include, but are not limited to, community-wide risk reduction projects; efforts to improve the resilience of critical infrastructure and key resource lifelines; risk reduction for specific vulnerabilities from natural hazards or acts of terrorism; and initiatives to reduce future risks after a disaster has occurred.

Natural Resources – Land, fish, wildlife, biota and water. Water means salt and fresh water, surface and ground water used for drinking, irrigation, aquaculture and recreational purposes, as well as in its capacity as fish and wildlife habitat.

NGO – A nongovernmental entity that serves the interests of its members, individuals, or institutions and is not for private benefit.

Recovery – Those capabilities necessary to assist communities affected by an incident to recover effectively, including, but not limited to, rebuilding infrastructure systems; providing adequate interim and long-term housing for survivors; restoring health, social, and community services; promoting economic development; and restoring natural and cultural resources.

Redevelopment – Rebuilding degraded, damaged or destroyed social, economic and physical infrastructure in a community, State or Tribal government to create the foundation for long-term development.

Resilience – Ability to adapt to changing conditions and withstand and rapidly recover from disruption due to emergencies.

Response – Those capabilities necessary to save lives, protect property and the environment, and meet basic human needs after an incident has occurred.

Restoration – Returning a physical structure, essential government or commercial services or a societal condition back to a former or normal state of use through repairs, rebuilding or reestablishment.

Short-Term Recovery – Phase of recovery which addresses the health and safety needs beyond rescue, the assessment of the scope of damages and needs, the restoration of basic infrastructure and the mobilization of recovery organizations and resources including restarting and/or restoring essential services for recovery decisionmaking.

Smart Planning – An urban planning and transportation theory that incorporates the concepts of smart growth and advocates for concentrating growth in the center of a city to avoid urban sprawl and promotes compact, transit-oriented, walkable, bicycle- friendly land use, including neighborhood schools, complete streets, and mixed-use development with a range of housing choices. It values long-range, regional considerations of sustainability over a short- term focus.

Stabilization – The process by which the immediate impacts of an incident on community systems are managed and contained.

Steady-State – A state where operations and procedures are normal and ongoing. Communities are considered to be at a steady-state prior to disasters and after recovery is complete.

Sustainability – Meeting the needs of the present without compromising the ability of future generations to meet their own needs.

Underserved Populations/Communities – Groups that have limited or no access to resources or that are otherwise disenfranchised. These groups may include people who are socioeconomically disadvantaged; people with limited English proficiency; geographically isolated or educationally disenfranchised people; people of color as well as those of ethnic and national origin minorities; women and children; individuals with disabilities and others with access and functional needs; and seniors.

Guide to Figures and Tables

Figures

Tables

13. APPENDICES

Appendix A. Additional resources

- Chapter 7 of the *Americans with Disabilities Act (ADA) Best Practices Tool Kit, Emergency Management* at: http://www.ada.gov/pcatoolkit/toolkitmain.htm
- *Comprehensive Planning Guide 101* at: http://www.fema.gov/about/divisions/cpg.shtm
- *Critical Infrastructure and Key Resources (CIKR)* at: http://www.dhs.gov/files/programs/gc_1189168948944.shtm
- *Disaster Assistance (DA)* at: http://www.disasterassistance.gov
- Economic Recovery Resources at: http://restoreyoureconomy.org
- *Executive Order 13175 – Consultation and Coordination with Indian Tribal Governments* at: http://ceq.hss.doe. gov/nepa/regs/eos/eo13175.html
- Title VI, *Rehabilitation Act of 1973* at: http://www.access-board.gov/enforcement/rehab-act- text/intro.htm

- *Age Discrimination Act of 1975* at: http://www.dol.gov/oasam/regs/statutes/age_act.htm
- *Fair Labor Standards Act* at: http://www.dol.gov/whd/flsa/index.htm
- *Hurricane Ike Impact Report: Special Needs Populations Impact Assessment Source Document, White Paper* at: http://www.disabilitypreparedness.gov/pdf/ike_snp.pdf
- *SME: Interagency Council on Emergency Preparedness and Individuals with Disabilities (ICC)* at: http://www.dhs.gov/files/committees/editorial_0591.shtm
- *National Commission on Children and Disasters: Interim Report* at: http://cybercemetery.unt.edu/archive/nccd/20110426214402/http://www.acf.hhs.gov/ohsepr/nccdreport/nccdreport.pdf
- *National Council on Disability (NCD) August 2009 Report: Effective Emergency Management: Making Improvements for Communities and People with Disabilities* at: http://www.ncd.gov/publications/2009/ Aug122009
- *National Disaster Housing Strategy (NDHS)* at: http://www.fema.gov/pdf/emergency/disasterhousing/NDHS-core.pdf
- *National Disaster Housing Task Force (NDHTF)* at: http://www.fema.gov/emergency/disasterhousing/national_task_force
- *National Disaster Recovery Program Database* at: https://asd.fema.gov/inter/ndhpd/public/home.htm
- *National Flood Insurance Program (NFIP)* at: http://www.fema.gov/business/nfip/
- *National Incident Management System (NI MS)* at: http://www.fema.gov/emergency/nims/
- *National Infrastructure Protection Plan (NIPP)* at: http://www.dhs.gov/nipp
- *National Labor Relations Board (NLRB)* at: http://www.nlrb.gov/
- *National Response Framework (NRF)* at: http://www.fema.gov/pdf/emergency/nrf/nrf-core.pdf
- *National Voluntary Organizations Active in Disaster (NVOAD), "Disaster Spiritual Care Points of Consensus"* at: http://www.nvoad.org/resource-library/documents/doc_download/12-disaster- spiritual-care
- *Occupational Safety and Health Administration (O S H A)* at: http://www.osha.gov/
- *Department of Transportation's (DOT) National Transportation Recovery Strategy* at: http://www.dot.gov/disaster_recovery/
- *Robert T. Stafford Disaster Relief and Emergency Assistance Act (Stafford Act) (Public Law 93-288)* as amended at: http://www.fema.gov/about/stafact.shtm

Appendix B. Recommended Roles and Activities

Individuals and Families Recommended Roles and Activities

Table 24. Individuals and Families Pre-Disaster Checklist

• Mitigate home vulnerabilities by adding hurricane shutters, bracing cripple walls, anchoring bookshelves, maintaining a defensible firebreak around the house and pruning overhanging tree limbs among other strategies.

• Develop an individual/family disaster preparedness and recovery plan, seeking assistance from service providers as necessary. See www.ready.gov. • Participate in ongoing community-wide planning initiatives, including those specifically focused on pre-disaster disaster preparedness, recovery and mitigation. • Provide community input on potential community disaster risks, potential impacts, and recovery planning through the use of voting, comment, organized efforts and other means. Community input should consider building codes, flood plain management, proposed developments and environmental and natural resources rule making initiatives. • Purchase and maintain an appropriate and adequate level of hazard and flood insurance. • Assist others in obtaining planning guidance and tools. • Maintain supplies of food, water and battery-powered communications devices.

Table 25. Individuals and Families Post-Disaster Checklist

• Implement individual and family recovery plans. • Reach out to others who may need assistance. • Participate in post-disaster community recovery planning if possible. • Establish metrics to evaluate recovery progress and achievement of disaster recovery objectives for individuals and families. • Rebuild safer and stronger.

Private Sector Recommended Roles and Activities

Table 26. Private Sector Pre-Disaster Checklist

• Build relationships with community emergency managers and other recovery officials to have an active voice in the recovery process. • Develop, test and implement business continuity and restoration plans. Take into account worker safety and health and potential employee unavailability or attrition due to a disaster. • Educate and train employees to implement mitigation measures and preparedness activities consistent with business continuity plans. • Use internal communications channels to inform employees about preparedness efforts for work that address individual and family needs. • Carry adequate insurance to rebuild damaged facilities and to survive a disruption of work. • Incorporate mitigation measures in design and construction. • Mitigate risks from disasters by relocating from hazardous areas, hardening facilities and elevating critical infrastructure. • Identify products, services and technical assistance that would be needed for recovery. Align these needs with business sector resources that can be available in a post-disaster environment. • Participate and assume a leadership role in local recovery planning; articulate anticipated needs in a disaster and assist in identifying resources available to support recovery.

Table 27. Private Sector Post-Disaster Checklist

• Implement business continuity plans. • Communicate status of operations and supply chains as well as restoration challenges and timelines to local, State, Tribal or Federal recovery managers. • When possible, support employees impacted by the disaster by providing critical information on the recovery process through accessible and multilingual internal communications efforts. • Provide volunteers, leaders, technical assistance, commodities and facilities as willing and able.

Table 27. (Continued)

• Form business recovery groups or task forces to assist one another and to communicate more effectively with government and community leaders. • Research available funding sources and types of funding; understand the application processes of assistance programs. • Assist small and local businesses in acquiring assistance. • Rebuild safer and stronger. • Establish metrics to evaluate recovery progress and the achievement of private sector disaster recovery objectives.

Nonprofit Sector Recommended Roles and Activities

Table 28. Nonprofit Sector Pre-Disaster Checklist

• Build relationships with community emergency managers and other recovery officials to have an active voice in the recovery process. • Co-host stakeholder workshops in various accessible locations in the community to determine priority recovery issues that are informed by affected neighborhoods. • Incorporate mitigation in the design and construction of places of employment and promote mitigation to employees. • Implement lessons learned from disaster efforts into the planning process for the State Voluntary Organizations Active in Disaster (VOAD). • Actively participate in local pre-disaster recovery planning, articulating resources and capabilities and establishing partnership and support linkages with local VOADs. • Provide training related to post-disaster activities implemented by the organization. • Educate clients on the importance of mitigation strategies. • Establish systems and processes for nonprofit organizations, government agencies and individuals in remote areas to request post-disaster assistance.

Table 29. Nonprofit Sector Post-Disaster Checklist

• Deliver recovery resources and support services to vulnerable and underserved groups, individuals and communities as necessary. • Provide emotional and psychological care; include training for caregivers. • Supply housing repair, reconstruction and rehabilitation services that comply with applicable building codes and standards, zoning regulations and design standards. • Communicate and coordinate needs and capabilities to local, State and Tribal authorities with the Voluntary Agency Liaison Specialists (VALS). • Participate in the post-disaster community planning process. • Promote partnerships among all nongovernmental organizations (NGOs) conducting disaster recovery work.' • Serve as subject matter experts (SMEs) on subjects based on agency experience – for example, offer techniques for the handling of unsolicited donated goods or unaffiliated volunteers. • Coordinate recovery programs and services with other entities involved in recovery including government emergency management officials to ensure a unified recovery process that maximizes effectiveness of the overall effort. • Establish metrics to evaluate recovery progress and the achievement of nonprofit disaster recovery objectives

Local Government Recommended Roles and Activities

Table 30. Local Government Pre-Disaster Checklist

• Lead local preparedness, pre-disaster recovery and mitigation planning. • Engage community mapping initiatives that visually depict or otherwise identify known vulnerable geographic areas and infrastructure systems, at-risk subpopulation groups, economically disadvantaged neighborhoods/communities, resource available areas, and projected post-disaster impacts. • Encourage individuals and families to prepare for their recovery. • Use internal communications channels to inform employees about preparedness efforts for work and that address individual and family needs. • Pre-identify a structure for managing recovery, including identifying duties of a Local Disaster Recovery Manager (LDRM) for managing recovery. • Establish agreements and mechanisms to address surge capacity needs. • Ensure plans, agreements and operational initiatives address the provision of disability-related assistance and functional needs support services. • Institute mechanisms for immediate post-disaster damage assessments (i.e., train community residents and business owners, recruit PDA volunteers, expand on Citizen Corps efforts) and develop a routine process for informing State officials about disaster impacts. • Ensure compliance with local, State and Federal civil rights obligations. • Develop building and accessibility codes and land use standards as well as enforcement mechanisms which can reduce vulnerability to future disasters.

Table 31. Local Government Post-Disaster Checklist

• Organize, develop, implement and modify recovery, mitigation and land use plans as needed. • Appoint Local Disaster Recovery Manager (LDRM) and define activities and duties. • Ensure integrated efforts across government offices, the private sector and nongovernmental organizations (NGOs) during the formulation and implementation phase of recovery projects and activities, including raising and leveraging recovery funds. • Lead efforts to restore and revitalize all sectors of the community, including local critical infrastructure and essential services, business retention and the redevelopment of housing units damaged, disrupted or destroyed by the disaster. • Manage rebuilding so that risk reduction opportunities are optimized and comply with standards for accessible design. • Communicate and coordinate with other levels of government involved in recovery. • Undertake an appropriate community planning process – see Section 10 "Community Considerations." • Establish metrics to evaluate and communicate progress and the achievement of local disaster recovery objectives to all populations.

State Government Recommended Roles and Activities

Table 32. State Government Pre-Disaster Checklist

• Implement the State recovery and mitigation plans to include *Continuity of Government (COG)* and *Continuity of Operations (COOP).* • Create and manage requirements and incentives for pre-incident disaster recovery

preparedness and planning as well as hazard mitigation actions. • Ensure that updated and FEMA approved mitigation plans are maintained at the State level. • Identify recovery activities that are either primarily the responsibilities of State government or beyond the capabilities and/or authorities of local governments. • Identify responsibilities for the position of a State Disaster Recovery Coordinator (SDRC) or equivalent, and resources for State recovery support functions. • Provide technical assistance and training to local governments and nongovernmental organizations (NGOs) on State plans, programs and other resources for disaster recovery. • Ensure that adequate staffing and expertise are available. • Establish agreements and mechanisms to address surge capacity needs. • Implement applicable laws and regulations to protect the rights of community members to ensure physical, programmatic and communications access to preparedness activities and services so that preparedness information for underserved populations is available and accessible. • Develop and aid enforcement of building and accessibility codes and land use standards, which can reduce vulnerability to future disasters. • Support local area efforts to conduct immediate damage assessments and share information regarding damages. • Form a State-led Disaster Housing Task Force to develop a disaster housing strategy that outlines potential approaches in response to specific disasters. • Develop State Recovery Support Function (RSF)-equivalent recovery framework that addresses housing, economic, environmental, infrastructure, and health and social services needs – at a minimum.

Table 33. State Government Post-Disaster Checklist

• Implement the State recovery and mitigation plan. • Activate the State Disaster Recovery Coordinator (SDRC). • Assess local government recovery needs and capacities for the specific incident and assist local governments and communities with identifying recovery resources. • Coordinate with local, Tribal and Federal governments and agencies, private businesses and nonprofit organizations to lead and coordinate State recovery planning and assistance to impacted communities. • Lead unified recovery efforts of State agencies, setting appropriate State policies to guide State agency activities as well as inform the application of Federal funding. • Receive, record and manage Federal grant resources; ensure efficient, nondiscriminatory and effective use of the funds; enforce accountability and legal compliance. • Oversee volunteer and donation management and coordinate with Voluntary Agency Liaison Specialists (VALS). • Facilitate and oversee an accessible and inclusive case management process. • Develop and implement strategies for raising and leveraging recovery funds through private investments, charitable donations and State sources such as emergency funds, taxes, fees and bonds that are within the State's authority to seek. • Provide timely and accessible public information and manage expectations, in coordination with local, Tribal and Federal stakeholders. • Enact new or existing exemptions to State laws and/or regulations to requirements that facilitate rebuilding activities and promote safer, stronger and smarter building. • Coordinate with Federal law enforcement to prosecute disaster-related fraud, waste, discrimination and abuse and recover lost funds. • Establish metrics in coordination with the impacted communities to evaluate recovery progress and the achievement of statewide disaster recovery objectives. • Ensure safety and health of State workers.

Tribal Government Recommended Roles and Activities

Table 34. Tribal Government Pre-Disaster Checklist

- Enhance cooperation and partnerships with local and State governments.
- Lead Tribal pre-disaster recovery and mitigation planning efforts to include *Continuity of Government (COG)* and *Continuity of Operations (COOP).*
- Preserve and protect cultural resources, sacred sites and traditional lands.
- Integrate the needs of individuals having functional needs into all planning efforts.
- Develop a Tribal disaster hazard mitigation plan.
- Facilitate communication between the Tribal government and U.S. Government by informing the latter of cultural differences, Tribal distinctions and best means for communicating within the Tribal hierarchy and reaching underserved populations.
- Prepare a pre-disaster plan that outlines responsibilities, allows for the creation of a Tribal Disaster Recovery Coordinator (TDRC) position or equivalent and includes an organizational structure to manage recovery assistance application and allocation.
- Encourage individuals and families to prepare for their recovery.
- Establish agreements and mechanisms to address surge capacity needs.
- Institute mechanisms for immediate post-disaster damage assessments (i.e., train community residents and business owners, recruit PDA volunteers, expand on Citizen Corps efforts) and develop a routine process for informing Federal officials about disaster impacts.
- Develop and aid enforcement of building and accessibility codes and land use standards, which can reduce vulnerability to future disasters.
- Form a Tribal-led Disaster Housing Task Force to develop a disaster housing strategy that outlines potential approaches in response to specific disasters.

Table 35. Tribal Government Post-Disaster Checklist

- Define the Tribal community's recovery goals.
- Partner with local, State, Tribal and Federal agencies to assess needs, resources and recovery capabilities.
- Appoint a Tribal Disaster Recovery Coordinator (TDRC) or equivalent, and establish an organizational structure to manage recovery assistance application and allocation.
- Provide timely and accessible public information to Tribal community members and manage expectations, in coordination with local, Tribal and Federal stakeholders.
- Participate in long-term recovery planning committees with local, State, Tribal and Federal partners.
- Coordinate with local, State, Tribal and Federal governments to expedite assistance.
- Update and implement pre-disaster recovery and mitigation plans.
- Implement a system to apply, receive and manage recovery grant resources unique to Tribal governments.
- Establish metrics to evaluate recovery progress and the achievement of Tribal disaster recovery objectives.
- Ensure integrated efforts across government offices, the private sector and nongovernmental organizations (NGOs) during the formulation and implementation phase of recovery projects and activities, including raising and leveraging recovery funds.
- Lead efforts to restore and revitalize all sectors of the community, including critical infrastructure and essential services, business retention and the redevelopment of housing units damaged, disrupted or destroyed by the disaster.
- Manage rebuilding so that risk reduction opportunities are optimized and comply with standards for accessible design.

Table 35. (Continued)

• Facilitate and oversee an accessible and inclusive case management process. • Enact new or existing exemptions to laws and/or regulations to facilitate rebuilding activities and promote safer, stronger and smarter building. • Ensure safety and health of Tribal government workers.

Federal Government Recommended Roles and Activities

Table 36. Federal Government Pre-Disaster Checklist

• Develop Federal Recovery Support Functions (RSFs) readiness for disaster recovery response. • Promote recovery preparedness by providing guidance to local, State and Tribal governments and nongovernmental organizations (NGOs) on pre-disaster recovery planning. • Encourage use of steady-state grant programs for pre-disaster recovery mitigation, planning and preparedness activities that comply with Federal civil rights laws. • Conduct recovery preparedness planning, training and exercises based on Federal agency roles and responsibilities in disaster recovery. • Whenever possible, offer incentives to grant recipients to incorporate hazard mitigation, sustainability and natural and cultural resource protection techniques in any land use or infrastructure projects. • Provide leadership for national catastrophic incident recovery planning. • Facilitate the sharing of planning best practices and recovery plans adopted by various local and State jurisdictions. • Conduct education and outreach for disaster recovery programs and resources to potential recipients and other stakeholders. • Foster a culture of open government by incorporating the values of transparency, participation and collaboration into programs, planning and daily operations to include underserved populations. • Evaluate program effectiveness; remove regulatory barriers, incorporate lessons learned and best practices while leveraging innovative technologies to increase efficiency and propose and enact reform changes as necessary. • Explore research on effective methods that can be used at the local, State and Tribal levels. • Ensure Federal agencies, where permissible, require mitigation activities as a criterion for approving funding to local, State and Tribal governments. • Conduct education and outreach for mitigation, disaster recovery programs and resources to potential recipients and other stakeholders

Table 37. Federal Government Post-Disaster Checklist

• Deploy a Federal Disaster Recovery Coordinator (FDRC) and activate and deploy Recovery Support Functions (RSFs) when determined necessary and establish a recovery coordination structure in close collaboration with affected local, State and Tribal governments. • Identify how Federal programs can effectively address and support recovery needs. • Provide timely, accurate and accessible information to the public and manage expectations in coordination with local, State, Tribal and other stakeholders. • Monitor and make necessary adjustments to Federal assistance programs and their delivery to more appropriately and timely address recovery needs of the affected local, State and Tribal communities. • Ensure transparency and accountability of Federal expenditures that aid disaster recovery.

- Coordinate with the various State agencies and officials to ensure that they have an understanding of how to avoid duplicate payments and whom to contact at the various Federal agencies to answer related questions.
- Participate in and support local, State and Tribal recovery planning and mitigation efforts through technical assistance, expertise or other assistance as requested and needed.
- Coordinate Federal recovery efforts with private and nonprofit organizations in cooperation with local, State and Tribal officials.
- Develop, or refine existing, metrics to evaluate recovery progress and the achievement of Federal disaster recovery objectives.

Appendix C. Planning for a Successful Disaster Recovery

Table 38. Individuals and Families Pre-Disaster Planning Activities

Individuals who prepare for their recovery help their community's recovery as well. Families and individuals are encouraged to:

- Develop an individual or family plan (that includes pets if appropriate) for disaster recovery.
- Become knowledgeable about hazards and hazard locations in and around home, work and the community.
- Self-assess risk exposure and incorporate mitigation practices.
- Purchase and maintain appropriate and adequate levels of insurance for potential hazards in the area.
- Work with others to prepare and reach out to those who may need assistance.
- Acquire and maintain survivability skills such as first aid.

Table 39. Individuals and Families Post-Disaster Planning Activities

Disaster recovery begins with individuals and families with an emphasis on personal responsibility in preparedness and recovery efforts. Individuals and families are encouraged to:

- Take charge of managing individual and family (including pets where appropriate) recovery.
- Participate in post-disaster community recovery planning.
- Work with others to reach out to those who may need assistance.
- Engage in public involvement opportunities on specific recovery projects.

Table 40. Private Sector/Businesses Pre-Disaster Planning Activities

Pre-disaster planning for the private sector and businesses integrates with community planning. The private sector and businesses are encouraged to:

- Identify and understand areas of risk.
- Develop, test and implement business continuity and restoration plans.
- Provide training for community business leaders to assist with the business recovery process.
- Participate in community pre-disaster planning, training and exercises.
- Incorporate hazard mitigation in the design and construction of places of employment and promote hazard mitigation to employees.

Table 40. (Continued)

• Build relationships with emergency managers to ensure an active voice in the recovery process. • Identify leaders and others to participate in community long-term recovery committees

Table 41. Private Sector/Businesses Post-Disaster Planning Activities

The private sector and businesses play an important role in the community and are often part of community leadership. The private sector and businesses are encouraged to: • Participate in local recovery planning. • Implement business continuity and restoration plans. • Assume a significant role in local and State recovery organizations.

Table 42. Nonprofit Sector Pre-Disaster Planning Activities

Nonprofits play a valuable role in communities and their preparedness efforts. Nonprofits are encouraged to: • Define and understand areas of risk. • Develop, test and implement business continuity and restoration plans. • Build relationships with community emergency managers to ensure an active voice in the recovery process. • Facilitate and encourage the participation of leaders and representatives from traditionally underserved populations in local long-term recovery committees and recovery planning workgroups. • Participate in community pre-disaster planning. • Incorporate hazard mitigation in the design and construction of places of employment and promote hazard mitigation to employees. • Identify resources to provide services in as equitable a manner as possible. • Determine pre-disaster planning and training opportunities related to post-disaster activities implemented by the organization. • Assist the local government with planning for the needs of individuals with disabilities and others with access and functional needs, individuals with limited English proficiency, children, seniors and other constituencies served by the nonprofit sector. • Help the local government with communicating disaster risks and vulnerabilities to individuals and families in an accessible and effective manner.

Table 43. Nonprofit Sector Post-Disaster Planning Activities

As important community partners, nonprofits are encouraged to: • Facilitate stakeholder workshops in various accessible locations in the community to determine priority issues for recovery that are informed by affected neighborhoods. • Implement business continuity and restoration plans. • Participate in local recovery planning. • Facilitate participation of leaders and representatives from traditionally underserved populations in local, State, Tribal and Federal recovery organizations. • Provide pre-disaster planning and training related to post-disaster activities implemented by the organization.

Table 44. Local Government Pre-Disaster Planning Activities

Local governments examine community-wide issues as part of pre-disaster planning. Local governments are encouraged to: • Understand key hazards, risks and vulnerabilities that cause systemic and major disruptions and challenges for disaster recovery, reconstruction and revitalization. • Communicate risks and vulnerabilities to the exposed community in an accessible and effective manner. • Include businesses and the nonprofit sector as partners in planning. • Pre-identify hazard mitigation goals, objectives and actions and incorporate them into ongoing pre-disaster recovery planning. • Incorporate hazard mitigation in design and construction and promote hazard mitigation to community members through measures including, but not limited to, the adoption and enforcement of appropriate building codes and standards. • Predetermine local recovery functions, roles, structures and funding for post-disaster recovery efforts to expedite the recovery process. • Determine how local disaster support functions work with State and Federal resources, to include Recovery Support Functions (RSFs). • Define critical infrastructure and key services that must be restored immediately post-disaster. • Maintain capability to expeditiously and effectively address recovery challenges, such as implementing building moratoriums, conducting damage assessments and issuing variances necessary to assist early recovery. • Plan for the needs of individuals with disabilities and others with access and functional needs, children and seniors as a fundamental aspect of the recovery plan rather than as a supplement or special plan. • Take Tribal law and culture into consideration in the community planning process; instill a respect and understanding for the unique heritage and needs of Tribal governments. • Work with government agencies to articulate and solidify collaborations between local, State and Tribal governments, particularly when multiple jurisdictions are involved and affected. • Identify community organizations with preestablished strategies in place to reach their stakeholders and coordinate information sharing with these organizations regarding planning activities and meetings. • Plan for the needs of individuals and families that have been displaced by the disaster. • Plan the coordination and outreach and awareness efforts to individuals with disabilities, individuals with limited English proficiency, seniors, children and other members of underserved populations.

Table 45. Local Government Post-Disaster Planning Activities

Recovery planning within a community is dependent on an active local government. Local governments are encouraged to: • Provide leadership in recovery planning and prioritization of goals. • Determine need and deploy a Local Disaster Recovery Manager (LDRM) or equivalent. • Incorporate principles of post-disaster planning into the recovery process. • Coordinate with relevant regional planning organizations that provide resources and/or planning expertise. • Promote partnerships between nonprofit organizations, faith-based organizations, the private sector or other relevant organizations and nontraditional and underserved populations throughout the recovery process.

Table 45. (Continued)

• Review pre-existing plans and cross-check against post-disaster planning priorities. • Implement a transparent, accountable system to manage recovery resources. • Manage overall recovery coordination at the local level. • Communicate post-disaster planning as well as organizational and operational needs to the State. • Lead an inclusive and accessible planning process, facilitating practices that comply with applicable laws, including civil rights mandates. • Enforce all applicable Federal worker protection laws for workers who are employed to rebuild the impacted community. These Federal laws include the *Fair Labor Standards Act, Occupational Safety and Health Regulations, National Labor Relations Act* and the laws administered by the Equal Employment Opportunity Commission (EEOC). Implement, coordinate and manage awareness and outreach efforts to individuals with disabilities, individuals with limited English proficiency, seniors, children, and other members of underserved populations

Table 46. State Government Pre-Disaster Planning Activities

To carry out their essential role in recovery, State governments are encouraged to: • Establish, organize and coordinate goals, objectives and timelines for recovery. • Plan and train for State Disaster Recovery Coordinator (SDRC) and Recovery Support Function (RSF) roles, structures and funding among State agencies and departments. • Create a post-disaster recovery authority for catastrophic-level incidents that operate immediately after a disaster and feature the legal and fiscal tools needed to ensure recovery. • Maintain a system to manage and monitor implementation of the recovery effort, enforce accountability, ensure accessibility and track resources. • Identify and encourage the use of specific standards and building codes to be used during reconstruction. • Ensure State laws and regulations do not inhibit effective recovery efforts. • Emphasize the importance of pre-disaster recovery planning at the local, State and Tribal levels. • Work with local governments to solidify collaborations between governments and to integrate pre-disaster recovery planning, such as response, land use and hazard mitigation planning, with capital improvement and other appropriate community planning for local governments. • Determine which organizations within the State and region have pre-established methods in place to reach their stakeholders. • Support local government efforts to identify organizations at the local level and coordinate information sharing with these organizations regarding planning activities and meetings. • Encourage local governments to establish an accessible and inclusive process for addressing recovery challenges pre- and post-disaster. • Include business and nonprofit sectors as partners in planning. • Coordinate and implement statewide hazard mitigation planning, projects and programs with recovery officials, disseminate hazard mitigation information and provide technical assistance for local mitigation efforts. • Update State hazard mitigation plans. • Provide a system of State level support to local governments that request assistance or lack capacity. • Communicate and coordinate with Federal recovery partners.

• Take Tribal law and culture into consideration in the community-planning process; instill respect and understanding for the unique heritage and needs of Tribal governments, as necessary. • Enter into agreements that articulate collaborations between local governments and Tribal governments, particularly when reservation land crosses multiple jurisdictions, as necessary. • Clarify relationships with both State and Federal authorities to determine where Tribal governments fit in the allocation of disaster resources when recovery initiatives begin. • Plan for the needs of individuals and families displaced by the disaster. • Pre-identify hazard mitigation goals, objectives and action and incorporate them into ongoing pre-disaster recovery planning. • Maintain capability to expeditiously and effectively address recovery challenges, such as conducting damage assessments and using variances necessary to assist early recovery. • Advise local communities and local private and nonprofit organizations on the coordination and outreach efforts to individuals with disabilities, individuals with limited English proficiency, seniors, children and other members of underserved populations.

Table 47. State Government Post-Disaster Planning Activities

States play an important role in supporting and, where necessary, leading overwhelmed local governments to address complex governmental, regulatory and financial challenges during short- and long-term recovery. State authorities: • Determine need and deploy State Disaster Recovery Coordinator (SDRC) or equivalent. • Provide a system of support to local governments. • Coordinate efforts to meet recovery challenges across all sectors in collaboration with recovery counterparts at all jurisdictional levels. • Conduct post-disaster planning and build on the foundation constructed during the pre-disaster planning phase, modifying it based on actual versus predicted risk and needs. • Develop an interface between State agencies and the Federal Government to streamline recovery funding at the local level. • Keep the public informed on all aspects of recovery. • Engage relevant regional planning organizations to provide resources and/or planning expertise. • Promote partnerships among nonprofit organizations, faith-based organizations, the private sector or other relevant organizations and nontraditional and underserved populations throughout the recovery process. • Implement and enforce applicable requirements to protect the rights of its community members needing physical, programmatic and communications access to recovery activities and services, and workers who are employed to rebuild the impacted community. • Support communities and nongovernmental organizations (NGOs) with coordination and outreach efforts to individuals with disabilities, individuals with limited English proficiency, seniors, children and other members of underserved populations.

Table 48. Tribal Government Pre-disaster planning activities

To accomplish pre-disaster planning Tribal governments are encouraged to: • Identify and understand areas of risk. • Preassign Tribal recovery functions, roles and responsibilities to include those of a Tribal Disaster Recovery Coordinator (TDRC) or equivalent.

Table 48. (Continued)

• Maintain a system to manage and monitor implementation of the recovery effort; enforce accountability, ensure accessibility and track resources. • Integrate recovery and hazard mitigation in community planning processes and encourage hazard mitigation measures. • Coordinate with local, State and Federal governments to facilitate post-disaster efforts and ensure Tribal governments have knowledge of and access to available funding and other assistance. • Work with local governments to articulate and solidify collaborations between Tribal and local governments, particularly when reservation land crosses multiple jurisdictions. • Include business and nonprofit sectors as partners in planning. Plan the coordination and outreach and awareness efforts to individuals with disabilities, individuals with limited English proficiency, seniors, children and other members of underserved populations.

Table 49. Tribal Government Post-disaster planning activities

Tribal governments may live on land that spans multiple jurisdictions. Coordination with those jurisdictions plays a key role in planning for a Tribe's recovery from a disaster. Tribal governments are encouraged to: • Determine need and deploy Tribal Disaster Recovery Coordinator (TDRC) or equivalent. • Establish, organize and coordinate goals, objectives and timelines for recovery. • Coordinate with local and State governments to provide mutual support. • Conduct post-disaster planning by building on the foundation constructed during the pre-disaster planning phase, modifying it based on actual versus predicted risk and needs. • Participate in long-term community recovery activities sponsored by State or neighboring local jurisdictions. • Develop a relationship with the Federal Government to clarify and streamline recovery funding. • Implement, coordinate and manage awareness and outreach efforts to individuals with disabilities, individuals with limited English proficiency, seniors, children and other members of underserved populations.

Table 50. Federal government pre-disaster planning activities

The Federal Government supplements post-disaster local, State and Tribal capability for short-, intermediate and long-term recovery governmental capacity planning and technical assistance after extraordinary large-scale disasters or catastrophic incidents. The Federal Government will: • Plan for national-level responsibilities of catastrophic and regional disaster recovery challenges. • Identify and train Federal Disaster Recovery Coordinators (FDRCs) and participants in Recovery Support Functions (RSFs) to prepare for activation as required. • Offer technical assistance to local, State and Tribal governments and stakeholders on the process, practices and policies of hazard mitigation. • Provide pre-disaster recovery and mitigation planning training and tools for local, State and Tribal governments.

- Optimize and coordinate Federal programs that support local, State and Tribal recovery efforts.
- Include businesses and the nonprofit sector as partners in planning.
- Communicate and coordinate with local, State and Tribal recovery partners.
- Encourage municipalities, States and Tribes that lack the resources to develop pre-disaster recovery plans to use steady-state grant programs to the extent consistent with the authorities, terms and conditions for those awards.
- Facilitate the sharing of planning best practices and recovery plans adopted by various Tribal governments, local and State jurisdictions.
- Identify and resolve potential conflicts among Federal programs or regulations that may impede timely recovery.
- Maintain capability to expeditiously and effectively address recovery challenges, such as conducting damage assessments and using variances necessary to assist early recovery.
- Support State and Tribal implementation of coordination and outreach efforts to individuals with disabilities, individuals with limited English proficiency, seniors, children and other members of underserved populations.

Table 51. Federal government post-disaster planning activities

The types of planning assistance and level of technical support available through the Federal Government varies by community needs and depends on disaster impacts and the recovery capacities of local, State and Tribal governments. The Federal Government will:

- Assess disaster recovery need and deploy a Federal Disaster Recovery Coordinator (FDRC) and appropriate Federal Recovery Support Functions (RSFs) if determined necessary.
- Ensure local ownership of the early recovery process through the engagement of local, State and Tribal authorities in the planning, execution and monitoring of recovery actions.
- Supplement, but not supplant local, State and Tribal resources, and support local, State and Tribal leadership of the recovery process.
- Coordinate with other Federal agencies (OFAs) to identify the geographic extent of disaster impacts and address the broader recovery challenges to infrastructure, supply chains, transportation systems and the like.
- Maintain a system for addressing intra-governmental recovery coordination.
- Assess the need for technical expertise to support prompt initiation of recovery and maximize joint recovery efforts and resources.
- Provide planning guidance, tools, resources and best practices to local, State and Tribal governments to facilitate their recovery planning.
- Implement regulations and guidance regarding legal obligations on all aspects of recovery, including applicable civil rights laws, such as those that pertain to accessibility standards and address the needs of individuals with disabilities and individuals with limited English proficiency.
- Develop an accessible public information campaign to increase stakeholder awareness of the processes involved in recovery.
- Support local, State, Tribal and Federal entities in the identification, use and management of Federal grants.
- Encourage the use of sustainable development and mitigation practices in disaster-affected areas.

End Notes

[1] The NDRF is not intended to, and does not, create any right or benefit, substantive or procedural, enforceable at law or in equity, by any party against the United States, its departments, agencies, or entities, its officers, employees, or agents, or any other person.

[2] For descriptions of the COG and COOP, see U.S. Department of Homeland Security, *Federal Continuity Directive 1 (FCD 1): Federal Executive Branch National Continuity Programs and Requirements*, http://www.fema.gov/pdf/about/offices/fcd1.pdf

[3] Note that the primary mission of the U.S. Department of Defense (DOD) and its components is national defense. Because of this critical role, resources are committed after approval by the Secretary of Defense or at the direction of the President. When Federal military and civilian personnel and resources are authorized to support civil authorities, command of those forces remain with the Secretary of Defense.

[4] Federal Emergency Management Agency (FEMA) Public Assistance, Category B Emergency Protective Measures, may assist in providing some essential health and social services temporarily, if eligible.

[5] Recovery requires a heavily-leaning forward strategy because there are typically more resources available during the response phase, and recovery needs could be potentially lessened or prevented by leveraging *Stafford Act* funds.

[6] Please see Section 3.3 for primary agencies designated for recovery-specific missions.

[7] *National Infrastructure Protection Plan (NIPP)*, 2009, identifies a *Partnership Framework* composed of 18 sectors. Designation of the sectors is derived from HSPD-7.

[8] In certain catastrophic situations and other extreme conditions, the Title I authorities of the *Defense Production Act* may be used to prioritize and allocate key resources and services in the interest of national security and defense and to support critical infrastructure restoration. For more information, see 50 U.S.C. App. § 2061 et seq. and http://www. fema.gov/about/programs/dpa/pubs.shtm and http://www.bis.doc.gov/dpas

[9] These elements modified from the *Long-Term Community Recovery (LTCR) Planning Process; A Self-Help Guide*. http://www.fema.gov/library/viewRecord.do?id=2151

INDEX

A

B

C

D

E

F

G

H

I

R

S

T

U

V

W

Y